WIRELESS SENSOR
AND ACTOR NETWORKS II

IFIP – The International Federation for Information Processing

IFIP was founded in 1960 under the auspices of UNESCO, following the First World Computer Congress held in Paris the previous year. An umbrella organization for societies working in information processing, IFIP's aim is two-fold: to support information processing within its member countries and to encourage technology transfer to developing nations. As its mission statement clearly states,

> *IFIP's mission is to be the leading, truly international, apolitical organization which encourages and assists in the development, exploitation and application of information technology for the benefit of all people.*

IFIP is a non-profitmaking organization, run almost solely by 2500 volunteers. It operates through a number of technical committees, which organize events and publications. IFIP's events range from an international congress to local seminars, but the most important are:

- The IFIP World Computer Congress, held every second year;
- Open conferences;
- Working conferences.

The flagship event is the IFIP World Computer Congress, at which both invited and contributed papers are presented. Contributed papers are rigorously refereed and the rejection rate is high.

As with the Congress, participation in the open conferences is open to all and papers may be invited or submitted. Again, submitted papers are stringently refereed.

The working conferences are structured differently. They are usually run by a working group and attendance is small and by invitation only. Their purpose is to create an atmosphere conducive to innovation and development. Refereeing is less rigorous and papers are subjected to extensive group discussion.

Publications arising from IFIP events vary. The papers presented at the IFIP World Computer Congress and at open conferences are published as conference proceedings, while the results of the working conferences are often published as collections of selected and edited papers.

Any national society whose primary activity is in information may apply to become a full member of IFIP, although full membership is restricted to one society per country. Full members are entitled to vote at the annual General Assembly, National societies preferring a less committed involvement may apply for associate or corresponding membership. Associate members enjoy the same benefits as full members, but without voting rights. Corresponding members are not represented in IFIP bodies. Affiliated membership is open to non-national societies, and individual and honorary membership schemes are also offered.

WIRELESS SENSOR AND ACTOR NETWORKS II

Proceedings of the 2008 IFIP Conference on Wireless Sensor and Actor Networks (WSAN 08), Ottawa, Ontario, Canada, July 14-15, 2008

Edited by
Ali Miri
University of Ottawa
Canada

 Springer

Editor
Ali Miri
University of Ottawa
Canada

p. cm. (IFIP International Federation for Information Processing, a Springer Series in Computer Science)

ISSN: 1571-5736 / 1861-2288 (Internet)
ISBN: 978-0-387-09440-3   e-ISBN: 978-0-387-09441-0

Library of Congress Control Number: 2007934347

Printed on acid-free paper

9 8 7 6 5 4 3 2 1

springer.com

Preface

The 2008 IFIP Conference on Wireless Sensors and Actor Networks (WSAN 08) is the second in a series of annual conferences which are dedicated to the development and application of Wireless and Actor Networks (WSAN). Initiated by the IFIP Working Group 6.8 Mobile and Wireless Communications, it was held last year at the Universidad de Castilla-La Mancha, Albacete, Spain. This years conference was held at the School of Information Technology and Engineering of the University of Ottawa, Ottawa, Canada. The School, located in one of Canada's largest high-technology centres, also includes a large number of researchers, and many research laboratories specializing in related research areas. The conference received submissions from 15 different countries. After a rigorous evaluation process by the program committee members, assisted by external reviewers, a total of 23 papers were selected to be included in the program. The program was organized into the following themes: *Energy, Actors, Security, MACs*, and *Protocols*.

I am grateful to the program committee members and the external reviewers for their hard work and expertise in selecting the program. I would like to thank the Paper Award Committee members for their assistance in selecting the best regular and student papers. I would also like to thank the organizing committee: the Publicity Co-Chairs, Pedro M. Ruiz and Behzad Malek; the Local Arrangement Chair, Ilker Onat; the Submission Chair, Xu Li; and the Publication Chair, Terasan Niyomsataya.

I hope that WSAN 2008 has been a memorable conference and an enjoyable experience for all of its participants.

Ali Miri
WSAN 2008 Chair

2008 IFIP Conference on Wireless Sensors and Actor Networks (WSAN 08)

July 14-15, 2008
School of Information Technology and Engineering,
University of Ottawa
Ottawa, Ontario, Canada

General Chair

Ali Miri	University of Ottawa, Canada

Technical Program Co-Chairs

Luis Orozco-Barbosa	Universidad de Castilla-La Mancha , Spain
Ivan Stojmenovic	University of Birmingham ,UK

Steering Committee

Augusto Casaca	INESC, Portugal
Pedro Cuenca	Chair IFIP WG 6.8
Otto Duarte	Universidad Federal de Rio de Janeiro, Brasil
Al Agha Khaldoun	University Paris-Sud, France
Pedro Marron	University of Bonn, Germany
Luis Orozco-Barbosa	Universidad de Castilla-La Mancha, Spain

Paper Award Committee

Silvia Giordano	DTI-SUPSI, Switzerland
Stephan Olariu	Old Dominion University, USA
David Simplot-Ryl	University Lille 1, France

Publicity Co-Chairs

Pedro M. Ruiz	Universidad de Murcia , Spain
Behzad Malek	University of Ottawa, Canada

Publications Chair

Terasan Niyomsataya	University of Ottawa, Canada

Submission Chair

Xu Li   Carleton University, Canada

Local Arrangement Chair

Ilker Onat   University of Ottawa, Canada

Technical Program Committee

Muneeb Ali   TU Delft, Netherlands
Guillermo Barrenetxea   EPFL, Switzerland
Maxim Batalin   UCLA, USA
Torsten Braun   University of Bern, Switzerland
Mehmet Ufuk Caglayan   Bogazii University, Turkey
Luigi Fratta   Politecnico di Milano, Italy
Hannes Frey   Southern Denmark University, Denmark
Lewis Girod   MIT, USA
Javier Gomez   UNAM, Mexico
Ali Grami   UoIT, Canada
Takahiro Hara   Osaka University, Japan
Mohamed Ibnkahla   Queen's University, Canada
Sassan Iraji   Nokia Research Center, Finland
Aman Kansal   Microsoft Research, USA
Srdjan Krco   Ericsson, Ireland
Jaime Lloret Mauri   University of Valencia, Spain
Hai Liu   University of Ottawa, Canada
Wei Lou   Polytechnic University, Hong Kong
Veljko Malbasa   University of Novi Sad, Serbia
Tommaso Melodia   University of Buffalo, USA
Sotiris Nikoletseas   CTI, Greece
Dan Steingart   UC Berkeley/Wireless Industrial Technologies,
                             USA
Kemal Ertugrul Tepe   University of Windsor, Canada
Fabrice Valois   INSA Lyon, France
Natalija Vlajic   York University, Canada
Vincent Wong   University of British Columbia, Canada
Zonghua Zhang   University Lille 1, France

Additional Reviewers

Markus Anwander   Alexander Gluhak   Miguel Lopez-Guerrero
Lei Ding   Athanassios Kinalis   Jia-Liang Lu
Klaus Doppler   Li-Chung Kuo   Georgios Mylonas

Michael Pascoe

Olivier Powell

Gerald Wagenknecht

Carl Wijting

Markus Wlchli

Tomoki Yoshihisa

Table of Contents

Energy

Actors

Security

Protocols

MAC And Protocols

Threat-Aware Clustering in Wireless Sensor Networks

Ryan E. Blace[1], Mohamed Eltoweissy[2] and Wael Abd-Almageed[3]

Abstract. Technological advances in miniaturization and wireless networking have enabled the utilization of distributed wireless sensor networks (WSN) in many applications. WSNs often use clustering as a means of achieving scalable and efficient communications. Cluster head nodes are of increased importance in these network topologies because they are both communication and coordination hubs. Much of the research into maximizing WSN longevity and efficiency focuses on dynamically clustering the network according to the residual energy contained within each node. This is a result of the commonly held assumption that battery depletion is the primary cause of node failure. In this work, we consider that there are applications in which threats may significantly impact node survival. In order to cope with these applications, we present a threat-aware clustering algorithm, extending the Hybrid Energy Efficient Distributed clustering algorithm (HEED) that minimizes the exposure of cluster heads to threats in the network environment. Simulation results indicate that our extended threat-aware HEED, or t-HEED, improves both the longevity and energy efficiency of a WSN while incurring minimal additional overhead. Our research demonstrates and motivates the need for a general framework for adaptive context-aware clustering in WSNs.

Keywords: context awareness, threat model, clustering, sensor networks

1. Introduction

Wireless sensor networks (WSN) are being employed in numerous environments and applications. Often, WSN implementations utilize clustering techniques as a method of achieving self-organization and scalability in their communication model [1]. WSNs are energy and resource constrained and their effectiveness and longevity is subject to that of their participant nodes.

Clustering techniques introduce heterogeneity to the service profile of the network, which has the side effect of creating nodes that can be considered 'more critical' than others. This is due to the fact that cluster head node serve as central hubs or super-peers for node management functions such as communications, organization, and security. In a clustered network, each sensor node forwards all communications toward the cluster head to which it is assigned. This can occur in one step or many steps, depending on the clustering implementation. It is the cluster head's job to route all received communications towards a destination or network sink. As a result, cluster heads are of increased importance to the proper functioning of the WSN. From this fact, it is apparent that the

---

[1] Ryan E. Blace, BBN Technologies LLC, Columbia, MD, reblace@bbn.com
[2] Mohamed Eltoweissy, Electrical and Computer Eng., Virginia Tech, toweissy@vt.edu
[3] Wael Abd-Almageed, UMIACS, University of Maryland, wamageed@umiacs.umd.edu

---

Please use the following format when citing this chapter:

Blace, R.E., Eltoweissy, M. and Abd-Almageed, W., 2008, in IFIP International Federation for Information Processing, Volume 264; Wireless Sensor and Actor Networks II; Ali Miri; (Boston: Springer), pp. 1–12.

compromise or loss of a cluster head will have a greater impact on the overall effectiveness and longevity of the network.

One of the primary design goals for clustering algorithms in WSNs is to maximize the lifespan of cluster heads [2-4]. By maximizing cluster head lifespan, one can minimize the need for the rearrangement of clusters, which is an expensive process involving, at a minimum, a number of communications between a cluster head and a sensor. Most research into clustering algorithms makes the assumption that the primary factor influencing the longevity of sensors in a WSN is the residual energy of the node.[3]. In other words, the primary cause of node failure is a depleted battery.

Increasingly, sensor networks are being used in environments where energy constraints are not the only threat to nodes. For example, in battlefield contexts, enemies may be actively searching for and destroying sensors. In the wilderness, firefighters may use air-deployed sensor networks to track the movements of wildfires that may overwhelm and destroy some of the nodes. In each of these contexts, it is likely that a node may be destroyed due to contextual factors other than energy levels.

This work explores the potential for incorporating a contextual threat-based element into an existing clustering technique that emphasizes energy conservation. Our hypothesis is that threat-aware clustering enhances network longevity and energy efficiency. As part of the work, we define a threat model and simulate the initial distribution and clustering of a sensor network using a clustering technique based on the Hybrid Energy Efficient Distributed clustering algorithm (HEED) [2], and our own proposed algorithm, t-HEED, which is an extension of HEED that uses contextual threat awareness to minimize the threat level of cluster heads. The network is then attacked by adversaries who perform the attacks based on a pre-determined threat distribution. We measured node count, residual energy, and other metrics after a number of epochs to assess the effectiveness of the contextual threat enhanced HEED implementation.

The remainder of the paper is organized as follows. Section 2 summarizes related work. Section 3 presents an overview of HEED [6] and describes our threat-aware clustering approach. Section 4 presents our network and threat models. Section 5 reports on our comparison between HEED and our extension, t-HEED. Finally section 6 concludes the paper and highlights future work.

2.   Related Work

Clustering has been demonstrated as an effective technique for achieving prolonged network lifetime and scalability in WSNs [5]. Parameters to include the node degree, transmission power, battery level, or processor load usually serve as metrics for choosing the optimal clustering structure. Recent initiatives address the problem of clustering and reclustering based on application specific attributes and network conditions. Bouhafs et al. [6] propose a semantic clustering algorithm for energy-efficient routing in WSNs. Nodes join the clusters depending on whether they satisfy a particular query inserted in the network. The output of the algorithm is called a semantic tree, which allows for layered data aggregation Siegemund [7] proposes a communication platform for smart objects that adapts the networking structure depending on the context. A cluster head node decides which nodes can join the cluster, based on similar symbolic location. Strohbach and Gellersen [8] propose an algorithm for grouping smart objects based on physical relationships. They use associations of the type "objects on the table" for constructing the clusters. A master node has to be able to detect the relationships for adding/deleting the nodes to/from the cluster. Perianu et al [9] propose a clustering scheme where the network is dynamic, the context is permanently changing and every pair of nodes is capable of understanding the physical relationships and thus the common context. Younis et al [10] propose dynamic cluster head relocation based on a tradeoff between safety and performance. A cluster head moves closer to an event for enhanced performance while considering the threat level along the path between the cluster head and the monitored event. Up to our knowledge, our work in this paper is the first treatment of threat-aware clustering.

3.   Threat-aware Clustering

To investigate the effectiveness of contextual threat-aware clustering, we modified HEED's clustering algorithm to consider a node's context when electing cluster heads. HEED is a distributed clustering algorithm that uses the residual energy of each node as a primary factor in deciding whether or not to become a cluster head. By introducing the threat weighting, our construction discourages nodes in high threat areas from becoming cluster heads. The next two subsections describe the HEED clustering algorithm and the modifications that were required to add context awareness.

3.1 HEED Clustering

The HEED clustering algorithm can be divided into three major steps: 1) Tentative cluster head distribution 2) Iterative CH election and balancing, and 3) Finalization and membership establishment. The algorithm is entirely distributed. All information must be transmitted between nodes, or known locally.

In the first step, each node decides whether or not to become a tentative cluster head based on a weighted probability of some ClusterHeadProbability*(Residual Energy/Max Energy). Essentially, each node has a fixed probability of becoming a cluster head, weighted by a dynamic measurement of the node's current residual energy. When a node elects to become a tentative cluster head, it broadcasts that information to all nodes within communication range.

In the second step, each node goes through an iterative process of deciding whether or not to become a final cluster head based on the cost of the nodes within its communication range and its own cluster head probability. Each node doubles its probability of becoming a cluster head after each successive iteration in which no decision is made. If a node determines it is the optimal cluster head, it will elect to become a tentative cluster head. Once a node's probability of becoming a cluster head has reached 1, it will assert itself as a final cluster head. Each node that is not 'covered' will repeat this process. A node becomes 'covered' when it is within communication range of either a final or tentative cluster head. Any time a node changes state during this phase, the node broadcasts the state change to all neighbors within communication range.

During the final phase of HEED clustering, each node decides to join the least cost cluster head. HEED uses one of two cost functions for determining cluster head membership, least degree and most degree. The goal of least degree is to balance load across all clusters. The goal of most degree is to provide dense clusters.

3.2   Threat-Aware HEED (t-HEED)

At first glance, the contextual threat HEED implementation appears to make minor modifications to the underlying HEED algorithm. While subtle, the changes significantly affect the clustering process.

First, we altered the initialization phase to assert the initial node distribution according to both the residual energy ratio and the contextual threat level. For our purposes, we assume that the local contextual threat level of each node can be objectively determined by each node. The new formula for cluster head probability is expressed in Equation (1). Since we are adding a new weight to the initial cluster head probability function, it may be necessary to increase the baseline cluster head probability C_{prob}. $E_{residual}$ and E_{max} refer to the current battery level and the maximum battery level of the node. $Threat_{prob}$ is the context-based threat value that defines the distribution of potential attacks. P_{min} is a minimum value that is needed so that CH_{prob} never becomes 0

$$CH_{prob} = \max\left(C_{prob} * \frac{E_{residual}}{E_{max}} * \left(1 - Threat_{prob}\right), P_{min} \right) \qquad (1)$$

Second, we altered the cost function that is used in both the Repeat phase and the Finalize phase of the algorithm. The cost function is used to rank the neighbors of a node according to their suitability as cluster heads. HEED provides the guidance of using either 'node degree' or '1/node degree' depending on whether the goal is to create dense clusters or to distribute load. We replace this with a simple measurement of the threat level of a node.

The final modification to HEED involves its determination of when a node should stop repeating the middle phase. HEED specifies that a node should repeat until it is 'covered'. A node is considered 'covered' if it is within communication range of a final or tentative cluster head. A node will stop trying to process its local maximum cluster head candidate at this phase because the presence of a tentative or final cluster head implies that the maximum is already known. This implies that the node can simply accept one of the already existing cluster head candidates as its cluster head.

In order for t-HEED implementation to properly propagate contextual information and determine the optimal solution, we had to loosen the definition of 'covered' to include only those nodes that have a final cluster head within communication range. Eliminating the tentative cluster heads from the list allows the algorithm to converge on a more optimal solution rather than be subject to the initial CH distribution which is based solely on local information.

In summary, the contextual threat information is used during each phase in order to modify the probability that a given node will become a cluster head based on its threat level. This has the effect of moving cluster heads away from high threat areas.

4   Network and Threat Models

4.1 Network Model

The simulated network model, Figure 1, is a simple, grid arranged, statically placed, sensor network. The simulations were performed with a grid size of 30 by 30 units. Nodes are distributed throughout the grid based on a normal random distribution. The simulations were performed with a node density of 0.35. It is assumed that nodes have a fixed communication range, simulated as 5 units, and communicate reliably. Each node contains a battery that is modeled as a float value initialized to 1. Each communication incurs a battery cost on both sender and receiver, simulated as 0.005 of the max battery capacity.

The network model simulates a clustered organization and communication scheme, as shown in Figure 2. During network operation, each node will be a cluster head or a sensor. Each sensor must be a member of exactly one cluster, attached to the cluster head by a maximum of one step. After deployment and clustering, any sensor that cannot reach a cluster head and is not a cluster head itself must turn itself off. Cluster heads should be placed such that they are uniformly distributed throughout the topology unless intentionally otherwise arranged.

For this simulation, I assume that after the initial clustering process, there is no reclustering. I am not attempting to test the effectiveness of HEED, simply to investigate the effects of adding the contextual threat awareness. In order to mitigate the potential problems associated with not supporting reclustering, the network model does provide node reassignment.

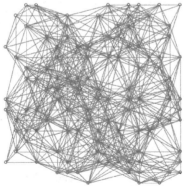

Fig. 1. Network Topology - initial distribution of nodes and node interconnectivity

The network model includes simulated traffic. Traffic is generated at a fixed rate and submitted to a random node in the network. If the source of the traffic is a cluster head, the traffic will incur a send cost on the CH. If the source of the traffic is a sensor node, a send cost is incurred on the sensor and a receive cost is incurred on the cluster head of the sensor. To clarify this point, the model simulates traffic that is routed from sensors to cluster heads. Cluster head to base station communication is ignored.

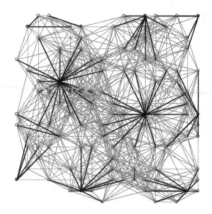

Fig. 2. Clustered organization of sensor nodes

When any node is killed either by a depleted battery or an adversary, it is turned off and effectively removed from the network. Sensor nodes that are killed have no effect on the other nodes in the network. Each cluster head that is killed has the effect of abandoning the nodes in its cluster. Abandoned nodes will attempt to reattach to other clusters within communication range. If no new cluster head is within range, the node will turn itself off.

4.2 Threat Model

The simulated threat model used against the network model can be characterized by a number of adversaries who travel fixed paths through the grid of nodes. Adversaries travel straight and consistent longitudinal and latitudinal paths through the network. Adversaries are generated at a fixed rate in bursts. At the start of the attack simulation, a number of adversaries are created and released. The adversaries travel their paths through the network, taking discrete steps of 0.1 units, until they have traveled across the network and traveled out the network boundaries. Once adversaries have left the network boundaries, the adversaries are destroyed. The attack simulation consistently maintains the number of adversaries until a specified max number of entities have traveled through the network. For the simulations performed in this work, adversaries were simulated at three different rates: 1 adversary at a time with a total of 10, 5 adversaries at a time with a total of 50, and 10 adversaries at a time with a total of 100.

When an adversary comes within a certain distance (simulated as 0.5 units) of a node in the network model, it probabilistically decides whether or not to attack the node. The probability is described in Equation (2). This probability ensures that an adversary is only going to attack a node once during its multi-step encounter with a node. The equation assumes that adversaries will travel directly over each node; however, the equation is a satisfactory way of maintaining the mode important probability: the attack success. The attacks of adversaries will succeed with a random probability that is according to the contextual threat level of the position that the adversary and node occupy. The probability is modeled in Equation (3).

$$P_{Attack} = \frac{1}{\left(\dfrac{KillDistance * 2}{StepSize} + 1 \right)} \quad (2)$$

$$P_{Success} \doteq P_{Attack} * Threat_{Prob} \quad (3)$$

When an adversary successfully attacks a node, whether it is a sensor or a cluster head, the node in question is immediately killed and cannot be revived. Adversary attacks are modeled as independent events. A successful attack does not alter the behavior of the adversary, it simply continues on its path.

The contextual threat probability distribution is modeled as a matrix of threat values that is 2 units larger in both dimension than the network grid. The distribution has two primary features: points and lines. Initially, the distribution is uniform, with each element in the matrix having a threat level of 0. First, a specified number of random point-based threats are established in the distribution, boosting the threat level of small, symmetric areas throughout the network model. Second, a specified number of random lines are established along latitude and longitude lines of the distribution. This simulates shape-based contextual threat elements (i.e. Roads or buildings). Two specific sets of parameter values were considered for simulation: a point-centric and a line-centric threat distributions. Simulation tests indicated that the distributions produced consistent results. As a result, the tests were performed against the line-centric model.

5   Simulation and Results

The simulation is developed in C# using the .Net platform 3.0. The simulation has a visual component and a date model component. The visual component is used to observe the simulation. Future efforts will include an implementation in TinyOS.

We ran extensive simulations using the network and threat models with the goal of gathering performance metrics from both the baseline HEED implementation and the extended threat-aware t-HEED implementation. The metrics are divided into two categories. The first category of metrics attempt to illustrate any effects the HEED modifications have on the behavior of the algorithm. These metrics include:

- The initial cluster head count
- The number of iterations required to converge on a clustering arrangement
- The average cluster head threat level

Ideally, the changes to HEED should not impact the initial cluster head count significantly. A significant change to cluster head count may bias other metrics and simulation results. The number of iterations required to converge on a cluster arrangement is important because it directly affects the time and energy required to perform clustering. Finally, the average cluster head threat level is an important metric that indicates how well the modifications are achieving their goal of minimizing cluster head threat exposure.

The second set of metrics capture the performance of the clustering technique by measuring the overall longevity and effectiveness of the WSN. These metrics include the number of nodes and cluster heads alive in the network, and the average residual energy of the nodes and cluster heads in the network. The number of nodes and cluster heads still alive in the network provide a simple indication of the remaining effectiveness of the network. The average residual energies of nodes and cluster heads in the network indicate the energy efficiency of the network and its potential longevity.

5.1 Clustering Metrics

Initial Cluster Head Count

Table 1 shows the initial cluster head count for two series of simulations. In the first series, all parameters were left at their regular levels and the density of the network was adjusted. In the second series, the transmission distance of each node in the network was adjusted. The results show that x HEED implementation created slightly fewer cluster heads in every test. The differences are marginal for the parameters that are closest to the defaults – Density 0.45 and transmission distance 6. An interesting observation to make is that the number of cluster heads increases with network density and decreases with transmission range.

Table 1: Initial Cluster Head Counts

t-HEED	HEED	
10.76	12.84	Density - 0.05
24.1	25.3	Density - 0.45
26.02	26.86	Density - 0.85
49.52	55.82	Trans – 3
16.58	16.84	Trans – 6
7.22	7.46	Trans – 10

Number of Iterations to Converge

Table 2 shows the number of iterations required to converge on a cluster arrangement for the same two series of simulations as for the initial cluster head count metric. In this case, t-HEED implementation took approximately double the iterations to converge than the baseline HEED implementation. Density does not appear to have significantly effected either implementation, and the transmission range seems to have an inverse relationship to convergence iterations.

Table 2: No. of Iterations before Convergence

t-HEED	HEED	
11.84	6	Density - .05
11.9	6	Density - .45

12	6	Density - .85
12	6	Trans – 3
11.6	6	Trans – 6
8.78	6	Trans – 10

Average Cluster Head Safety

Table 3 shows the average threat level of each cluster head in the WSN after clustering. The metric was measured for the same set of simulations as the other metrics. This is an important metric because it illustrates the effectiveness of the modified clustering algorithm at reducing the threat level of cluster heads. The numbers in the table are measurements of cluster head safety. The threat level is actually (1-x) where x is the value in the table. The data shows that t-HEED implementation is effective at reducing the threat level of cluster heads in the network.

Table 3: Average Threat Level for each Cluster

t-HEED	HEED	
0.6856	0.4555	Density - .05
0.7082	0.4622	Density - .45
0.7039	0.47	Density - .85
0.6472	0.4591	Trans – 3
0.7298	0.445	Trans – 6
0.7832	0.46	Trans – 10

Using the visualization tool, it is easy to see how the context extended implementation of HEED has performed (Figure 3) The cluster heads have moved into the low threat areas and are connected to sensor nodes that are in the high threat areas.

5.2 WSN Performance Metrics

Each implementation was subject to a series of simulations that measured the sensor and cluster head longevity and residual energy. The simulations were performed with the default parameters, except for the adversary count, which was tested at 1 at a time for a total of 10, 5 at a time for a total of 50 and 10 at a time for a total of 100 (see the section on the threat model for details). Each simulation was run 100 times and all values for all metrics are the average values over the course of the trials.

Figure 4 shows the number of sensors that remain alive at a given epoch. As the simulation runs, and adversaries attack the network, the number of live sensors decreases as expected. Figure 5 shows the same metric, only for cluster head nodes. The legends are the same for both diagrams. As observed, the benefits that t-HEED introduces are greater as the number of adversaries increases.

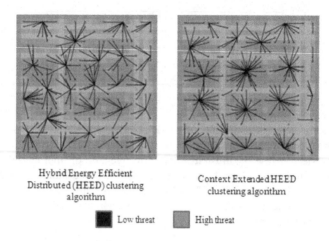

Hybrid Energy Efficient
Distributed (HEED) clustering
algorithm

Context Extended HEED
clustering algorithm

■ Low threat   ▨ High threat

Fig. 3. Comparison between HEED and threat-extended t-HEED clustering

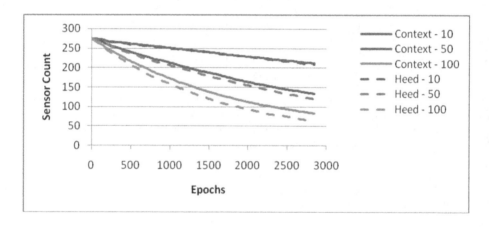

Fig.4. WSN Sensor Count

Figure 6 shows the residual energy for the same simulation. The source of the shift in the graph is uncertain, although it indicates that HEED uses more energy during the initial clustering phase than the context enhanced approach. Figure 7 shows the average residual energy for cluster head nodes in the network. The graph looks very similar to the alive node graph. This is due to the fact that the two metrics are closely correlated.

Additional simulations were performed with various adjusted parameters, including disabling node reattachment, disabling traffic generation, and tweaking node transmission range, network density, and initial cluster head probability. All of the results were consistent with those that have been presented. In every simulation, the context enhanced t-HEED implementation matched or outperformed the baseline HEED implementation.

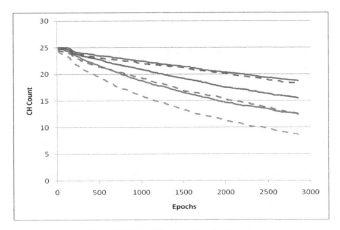

Fig. 5.WSN Cluster Head Count

6. Conclusion

We proposed threat-aware clustering to minimize the risk to cluster heads. The results show that using this approach increases the efficiency and longevity of a WSN without incurring an increased amount of overhead on the network. The primary conclusion is that context is a suitable parameter to be included in a clustering algorithm. Context need not be limited to a threat profile. The context awareness can extend to include expected locations of traffic generating events, or performance affecting factors like interference. Future work will include extending our proposed framework to a general framework for adaptive clustering based on context changes. Also, implementation Tiny OS and TOSSIM is a necessary future step in developing this work.

Acknowledgement
This work is supported in part by the National Science Foundation award No. 0721523.

References
[1] A. A. Abbasi and M. Younis, "A survey on clustering algorithms for wireless sensor networks." vol. 30: Butterworth-Heinemann, 2007, pp. 2826-2841.

[2] O. Younis and S. Fahmy, "HEED: A Hybrid, Energy-Efficient, Distributed Clustering Approach for Ad Hoc Sensor Networks," vol. 3, pp. 366-379, 2004.

[3] S. Yi, J. Heo, Y. Cho, and J. Hong, "PEACH: Power-efficient and adaptive clustering hierarchy protocol for wireless sensor networks." vol. 30: Butterworth-Heinemann, 2007, pp. 2842-2852.

[4] W.-T. Su, K.-M. Chang, and Y.-H. Kuo, "eHIP: An energy-efficient hybrid intrusion prohibition system for cluster-based wireless sensor networks." vol. 51: Elsevier North-Holland, Inc., 2007, pp. 1151-1168.

[5] I. Stojmenovic, M. Seddigh, and J. Zunic, "Dominating sets and neighbor elimination-based broad-casting algorithms in wireless networks," IEEE Transactions on Parallel and Distributed Systems, 13(1): 2002, pp.14–25.

6] F. Bouhafs, M. Merabti, and H. Mokhtar, "A semantic clustering routing protocol for wireless sensor networks," Consumer Communications and Networking Conference, IEEE Computer Society, 2006, pp. 351– 355.

[7] F. Siegemund, "A context-aware communication platform for smart objects." Pervasive, Elsevier, 2004, pp.69–86.

[8] M. Strohbach and H. Gellersen, "Smart clustering - networking smart objects based on their physical relationships," Proceedings of the 5th IEEE Int'l Workshop on Networked Appliances, IEEE Computer Society, 2002, pp. 151– 155.

[9] R.M Perianu, C. Lombriser, P. Havinga, J Scholten, G. Tröster, "Tandem: A Context-Aware Method for Spontaneous Clustering of Dynamic Wireless Sensor Nodes," Internet of Things, Int'l Conf. for Industry and Academia, March 2008.

[10] M. Younis, W. Youssef, M. Eltoweissy, and S. Olariu, "Safety- and QoS-Aware Management of Heterogeneous Sensor Networks," Journal of Inter- connection Networks, Vol. 7, No. 1, 2006, pp. 179-193.

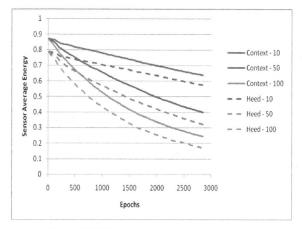

Fig. 6. WSN Sensor Node Average Energy

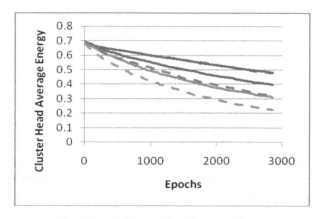

Fig. 7. WSN Cluster Head Average Energy

CES: Cluster-based Energy-efficient Scheme for Mobile Wireless Sensor Networks

Mohamed Lehsaini[1,2] , Hervé Guyennet[1], and Mohammed Feham[2]

[1] Laboratory of Computer Engineering of Franche-Comté University (LIFC),
16, Route de Gray, Besançon 25030, France
{lehsaini, guyennet}@lifc.univ-fcomte.fr
[2] Faculty of Engineering Sciences, Tlemcen University,
m_feham@mail.univ-tlemcen.dz

Abstract. Prolonging network lifetime has become a real challenge in Mobile Wireless Sensor Networks (MWSNs) as sensors have limited energy. In this paper, we propose a Cluster-based Energy-efficient Scheme (CES) for electing a cluster-head to evenly distribute energy consumption in the overall network and therefore obtain a longer network lifetime. In CES, each sensor calculates its weight based on k-density, residual energy and mobility and then broadcasts it to its 2-hop neighborhood. The sensor node with the greatest weight in its 2-hop neighborhood will become the cluster-head and its neighboring sensors will then join it. We performed simulations to illustrate the effects of sensor mobility on LEACH and LEACH-C's performance. Unfortunately, our findings showed that sensor mobility had a significant impact on both protocols' performance, but CES provided good results in terms of the amount of data packets received at the sink when compared with LEACH and LEACH-C.

Keywords: Cluster-head, k-density, Network lifetime, Residual energy, MWSNs.

1 Introduction

MWSNs consist of a large number of tiny mobile sensors that are randomly deployed in an interest area to sense phenomena. These mobile sensors collaborate with each other to form a sensor network able to send sensed phenomenon to a data collection point called the sink or base station. MWSNs could become increasingly useful in a variety of potential civil and military applications, such as intrusion detection, habitat and other environmental monitoring, disaster recovery, hazard and structural monitoring, traffic control, inventory management in factory environments and health related applications, etc. [1,2]. However, the deployment of MWSNs still requires solutions to a number of technical challenges that stem primarily from the constraints imposed by simple sensor devices: small storage capacity, low processing power, limited battery lifetime and short radio ranges.

Gathering information in MWSNs while minimizing the overall energy consumption and maximizing the amount of data received at the base station requires an efficient energy-saving scheme. Cluster-based architecture is considered an

Please use the following format when citing this chapter:

Lehsaini, M., Guyennet, H. and Feham, M., 2008, in IFIP International Federation for Information Processing, Volume 264; Wireless Sensor and Actor Networks II; Ali Miri; (Boston: Springer), pp. 13–24.

efficient approach to achieving this. Hence, we should imply determining parameters enabling to generate a reduced number of stable and balanced clusters.

The above constraints imposed by sensors make the design of an efficient scheme for prolonging MWSNs' lifetime a real challenge. In response to this challenge, we propose a Cluster-based Energy-efficient Scheme (CES) for MWSNs, which consists of grouping sensors into a set of disjoint clusters. In CES, the sensor with the greatest weight in its 2-hop neighborhood becomes the cluster-head. The weight of each sensor is calculated according to the following parameters: 2-density, residual energy and mobility. Furthermore, the cluster size ranges between two thresholds, $Thresh_{Lower}$ and $Thresh_{Upper,}$ which respectively represent the minimal and maximal number of sensors in a cluster. These thresholds are chosen arbitrarily or depend on network topology. Inside a cluster, each sensor is, at most, two hops from its corresponding cluster-head contrary to LEACH [3] and its variant LEACH-C [4], which allow only single-hop clusters to be constructed.

In the cluster-based heuristic methods proposed for WSNs, cluster members do not transmit their gathered data directly to the sink but to their respective cluster-head. Accordingly, cluster-heads are responsible for coordinating the cluster members, aggregating their sensed data, and transmitting the aggregated data to the remote sink, directly or via multi-hop transmission mode. Since cluster-heads receive many packets and consume a lot of power for long-range transmission, they are the ones whose energy is used up most quickly in the cluster if they are elected for a long time. Therefore, a cluster-based scheme should avoid a fixed cluster-head election scheme, because the latter has constrained energy and may rapidly drain its battery power due to heavy utilization. That can cause bottleneck failures in its cluster and trigger the cluster-head election process again. For that, we foresaw in the CES scheme that the cluster-head election process would be periodically carried out after a period of time called "round" to evenly balance the energy load among the sensors during the network lifetime.

In this paper we aim to minimize the energy consumption of the entire network and prolong the network lifetime. For this, we propose the CES scheme, which involves k-density and mobility factors in nodes' weight computation in order to guarantee the stability of clusters, as well as the energy factor to ensure a long cluster-head lifetime.

In our experiments, we conducted extensive simulations to evaluate the performance of both protocols: LEACH and LEACH-C with the same scenario presented in [3,4] but with mobile sensors. We also carried out simulations to evaluate CES's performance and compare the results obtained with LEACH and LEACH-C in terms of the amount of data packets received at the sink during the network lifetime.

The rest of this paper is organized as follows: in Section 2, we provide the necessary preliminary information for describing our scheme; Section 3 reviews several cluster-based algorithms that have been previously proposed; in Section 4, we present our new weighted scheme; and Section 5 presents a performance analysis of the proposed scheme. Finally, we conclude our paper and discuss future research work in Section 6.

2 Notations and hypothesis

Before heading into the technical details of our contribution, we shall start by giving some definitions and notations that will be used later in our paper.

A wireless sensor network is abstracted as an undirected graph $G=(V,E)$, called a connectivity graph, where V represents the set of wireless nodes and $E \subseteq V^2$ is the set of edges that gives the available communications; an edge $e=(u,v)$ belongs to E if and only if the node u is physically able to transmit messages to v and vice versa. Each sensor $u \in V$ is assigned a unique value to be used as an identifier so that the identifier of u is denoted by $Node_{Id}(u)$. The neighborhood set $N_1(u)$ of a node u is in (1). The size of this set is known as the degree of u, denoted by $\delta_1(u)$. The density of the network represents the average of the nodes' degrees.

$$N_1(u) = \{v \in V | v \neq u \wedge (u, v) \in E\}. \tag{1}$$

The 2-hop neighborhood set of a node u, i.e. the nodes which are the neighbors of u's neighbors except those that are u's neighbors, is represented by $N_2(u)$.

$$N_2(u) = \{w \in V | (v, w) \in E \text{ where } w \neq u \wedge w \notin N_1(u) \wedge (u, v) \in E\}. \tag{2}$$

The combined set of one-hop and two-hop neighbors of u is denoted by $N_{12}(u)$.

$$N_{12}(u) = N_1(u) \cup N_2(u). \tag{3}$$

In a general manner, the k-hop neighborhood set of a node u is represented by $N^k(u)$ as shown in (4) and its closet set of k-hop neighbors is denoted by $N^k[u]$ as in (5). Here, $d(u,v)$ represents the minimal distance in the number of hops from u to v. The size of $N^k(u)$ is known as the k-degree of u and denoted by $\delta^k(u)$.

$$N^k(u) = \{v \in V | v \neq u \wedge d(u, v) \leq k\} \tag{4}$$

$$N^k[u] = N^k(u) \cup \{u\} \tag{5}$$

The k-density of a node u represents the ratio between the number of links in its k-hop neighborhood (links between u and its neighbors and links between two k-hop neighbors of u) and the k-degree of u; formally, it is represented by the following formula:

$$k - density(u) = \frac{\left|(v, w) \in E : v, w \in N^k[u]\right|}{\delta^k(u)} \tag{6}$$

However, we are interested only in calculating the 2-density nodes so as not to weaken the CES scheme's performance as presented in (7). Table.1 illustrates the 2-density calculation of the nodes composing the network presented in Fig.1.

$$2 - \text{density(u)} = \frac{\left|(v, w) \in E : v, w \in N_{12}[u]\right|}{\delta^2(u)} \tag{7}$$

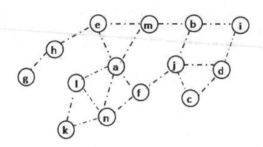

Fig. 1. Example of an abstracted wireless network

Table 1. Calculation of the nodes' 2-density.

Node	a	B	c	d	e	F	g	h	i	j	K	l	M	n
1-density	1,60	1	1,66	1,33	1,33	1,33	1	1	1	1,25	1,66	1,66	1,33	1,75
2-density	1,55	1,50	1,40	1,40	1,37	1,60	1	1,25	1,40	1,50	1,75	1,60	1,44	1,57

We propose to generate balanced clusters whose size ranges between two thresholds: $\text{Thresh}_{\text{Upper}}$ and $\text{Thresh}_{\text{Lower}}$. These thresholds are chosen arbitrarily or depend on network topology. If their values depend on network topology, they will be calculated as follows:

- u: the node that has the maximum number of 2-hop neighbors,

$$|N_{12}(u)| = \text{Max}(|N_{12}(u_i)| : u_i \in V) \tag{8}$$

- v: the node that has the minimum number of 2-hop neighbors,

$$|N_{12}(v)| = \text{Min}(|N_{12}(v_i)| : v_i \in V) \tag{9}$$

- Avg: the average number of 2-hop neighbors of all nodes in the network,

$$\text{Avg} = \frac{\sum_{i=1}^{n} N_{12}(u_i)}{n} \quad \text{where n : number of nodes} \tag{10}$$

$$\text{Thresh}_{\text{Upper}} = \frac{1}{2}(|N_{12}(u)| + \text{Avg}) \tag{11}$$

$$\text{Thresh}_{\text{Lower}} = \frac{1}{2}(|N_{12}(v)| + \text{Avg}) \tag{12}$$

In this paper, we assume that all sensors are given in a two dimensional space and we measure the distance between the two nodes u and v in terms of the number of hops. Each sensor has an omni-directional antenna which means that a single transmission from a sensor can be received by all sensors within its vicinity, and we consider that the sensors are almost stable in a reasonable period of time during the clustering process. We also assume that each sensor has a generic weight and that it is able to evaluate it. Weight represents the fitness of each node to be a cluster-head, and a greater weight means higher priority.

3 Related Work

Recently, many cluster-based techniques [3-12] have been proposed to deal with the main challenges in WSNs. However, most of these contributions focus on lifetime maximization in WSNs with stationary sensors. To the best of our knowledge, this paper is the first to tackle lifetime extension in WSNs with mobile sensors. In this section, we will review some of the most relevant papers related to cluster-based network architecture, which have been carried out to prolong lifetime in WSNs.

In [3], the authors propose LEACH, which is a distributed, single hop clustering algorithm for homogeneous WSNs. In LEACH, the cluster-head role is periodically rotated among the sensors to evenly distribute energy dissipation. After each round, each sensor elects itself as cluster-head with a probability which is equal to:

$$p_{CH} = k\frac{E(u)}{E_{Total}} \tag{13}$$

where E(u) represents remaining energy of node u, E_{Total} is the total energy in the whole network and k is the optimal number of clusters. However, the evaluation of E_{Total} presents a certain difficulty since LEACH operates without other routing schemes and any central control.

In [5], the authors compared homogeneous and heterogeneous networks in terms of energy dissipation in the whole network and analyzed both single-hop and multi-hop networks' performance. They chose LEACH as a representative of a homogeneous network and compared it with a heterogeneous single-hop network. The authors noticed that using single-hop communication between cluster members and their corresponding cluster-head may not be the best choice when the propagation loss index k (k>2) for intra-cluster communication is large, because LEACH might generate clusters whose size is important in dense networks and clusters whose size is limited in small networks. In both cases, cluster-heads could rapidly exhaust their battery power either when they coordinate among their cluster members or when they are placed away from the base station. Therefore, the authors proposed an improved version of LEACH called M-LEACH [5] (Muti-hop LEACH), in which cluster members can be more than one hop from their corresponding cluster-head and communicate with it in multi-hop mode. They also illustrate the cases where LEACH-M outperforms LEACH protocol. However, this proposed version requires each sensor to be capable of aggregating data, which increases the overhead for all sensors. To improve the performance of this strategy, in [6], the authors focus on heterogeneous sensor networks, in which two types of sensors are deployed: super and basic sensors. Super sensors have more communication and processing capabilities and act as cluster-heads, while basic sensors are simple (with limited power) and are affiliated to a nearby cluster-head and communicate with it directly or via multi-hop mode.

Furthermore, another variant of LEACH called LEACH-C [4] has been conceived to improve LEACH performance. This variant utilizes a centralized architecture to select cluster-heads while using a base station and location information from sensors. However, it increases network overhead since all sensors send their location information to the base station at the same time during every set-up phase. Several works have proven that a centralized architecture is particularly suitable for small networks, whereas it lacks scalability to handle the load when the network's size increases.

Similarly to LEACH-C, BCDCP (Base-Station Controlled Dynamic Clustering Protocol) [7] uses energy information sent by all sensors to the base station to build balanced clusters during the set-up phase. In BCDCP, the base station randomly changes cluster-heads while guaranteeing a uniform distribution of their locations in the interest field and carries out an iterative cluster splitting algorithm to find the optimal number of clusters. After that, it constructs multiple cluster-to-cluster (CH-to-CH) routing paths to use for data transfer, creates a schedule for each cluster and broadcasts it to the sensor network. In the second phase, which relates to data transfer, cluster-heads transmit collected data to the base station through the CH-to-CH routing paths [8]. However, BCDCP presents the same limitations as LEACH-C since it utilizes a centralized architecture to elect cluster-heads.

4 Our Contribution

In our proposed scheme, each sensor uses weight criteria to elect a cluster-head in its 2-hop neighborhood. The CES scheme assumes that sensors have 2-hop knowledge and operate asynchronously without a centralized controller. In CES, each sensor calculates its weight based on its k-density, its residual energy, and its mobility and broadcasts it to its 2-hop neighborhood. The sensor with the greatest weight in its 2-hop neighborhood is chosen as the cluster-head for the current round.

4.1 Cluster formation

The cluster formation process consists of grouping sensors into disjoint clusters, thus giving the network a hierarchical organization. Each cluster has a cluster-head which is chosen from its 2-hop neighborhood based on nodes' weight. The weight of each sensor is a combination of k-density, residual energy and mobility as presented in

(14), wherein the coefficient of each parameter can be chosen depending on the application.

$$\text{Weight}(u) = \alpha * 2 - \text{density}(u) + \beta * \text{Res} - \text{Energie}(u) + \gamma * \text{Mobility}(u) \quad (14)$$

$$\text{where } \alpha + \beta + \gamma = 1$$

Since the cluster head is responsible for carrying out several tasks - such as coordinating the cluster members, transmitting gathered data to the remote base station, and managing its own cluster - we propose to set up periodical cluster-head election processes after each round so that cluster-heads do not rapidly exhaust their battery power. We also propose that each cluster has a size ranging between two thresholds, $Thresh_{Lower}$ and $Thresh_{Upper}$, and that cluster members are, at most, 2-hops from their respective cluster-head.

In the CES scheme, each sensor is identified by a state vector as follows: $(Node_{Id}, Node_{CH}, Weight, Hop, Size, Thresh_{Lower}, Thresh_{Upper})$ where $Node_{Id}$ is the sensor identifier, $Node_{CH}$ represents the identifier of its cluster-head, Hop indicates the number of hops separating it from its respective cluster-head, and $Size$ represents the size of the cluster to which it belongs. Each sensor is responsible for maintaining a table called '$Table_{Cluster}$', in which information from the local cluster members is stored. The format of this table is defined as $Table_{Cluster}(Node_{Id}, Node_{CH}, Weight)$. The sensors could coordinate and collaborate between each other to construct and update the above stated table by using Hello messages. We used Hello messages to achieve these operations in order to alleviate the broadcast overhead and not degrade the CES scheme's performance. Moreover, each cluster-head has another table called '$Table_{CH}$', in which information from cluster-heads is stored. The format of this table is defined as $Table_{CH}(Node_{CH}, Weight)$.

Cluster formation is performed in two consecutive phases: set-up and re-affiliation.

4.1.1 The set-up phase

At the beginning of each round, each sensor calculates its weight and generates a 'Hello' message with two extra fields in addition to other regular contents: $Weight$ and $Node_{CH}$, where $Node_{CH}$ is set to zero. Then, it broadcasts it to its 2-hop neighborhood and eavesdrops on its neighbors' 'Hello' messages. The sensor with the greatest weight among its 2-hop neighborhood is chosen as the cluster-head (CH) for the current round. The latter updates its state vector by assigning the value of its identifier $Node_{Id}$ to $Node_{CH}$, and sets, respectively, Hop and $Size$ to 0 and 1. Then, it broadcasts an advertisement message (ADV_CH) including its state vector to its 2-hop neighborhood requesting them to join it, as illustrated by Fig. 2. Each sensor in the 1-hop neighborhood that receives the message and does not belong to any cluster and that has a lower weight than CH's weight, transmits a REQ_JOIN message to CH to join it. The corresponding cluster-head checks and, if its own cluster size does not reach $Thresh_{Upper}$, it will transmit an $ACCEPT\_CH$ message to this sensor; if not, it will simply drop the affiliation request message. Thereafter, CH increments its $Size$ value, and the affiliated sensor node sets Hop value to 1 and $Node_{CH}$ with $Node_{CH}$ as its corresponding cluster-head. Then, the affiliated sensors whose *Hop* value is equal to 1, broadcast received message again with the same transmission power to its neighbors. Similarly, each sensor belonging to $N_2(Node_{CH})$ that is not affiliated to any cluster and whose weight is lower than that of *CH*, transmits a *REQ_JOIN* message to the corresponding *CH*. In the same way, *CH* checks if its *Size* value remains under $Thresh_{Upper}$, and if so transmits ACCEPT_CH and updates its state vector. If not, it will drop the message of affiliation request. In the end, each sensor will know which cluster it belongs to and which sensor is its cluster-head.

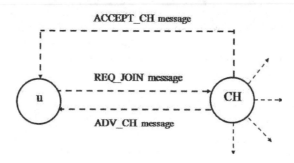

Fig. 2. Affiliation procedure of a node to a cluster

4.1.2 The re-affiliation phase

During the set-up phase, it may not be possible for all clusters to reach the $Thresh_{Upper}$ threshold. Moreover, it is possible that clusters whose size is lower than $Thresh_{Lower}$ may be created, since there is no constraint relating to the generation of these types of clusters. In this phase, we propose to re-affiliate the sensors belonging to clusters that have not attained the cluster size $Thresh_{Lower}$ to those that did not reach $Thresh_{Upper}$ in order to reduce the number of clusters formed and obtain balanced clusters.

The execution of this phase proceeds in the following way: cluster-heads that belong to clusters whose size is strictly lower than $Thresh_{Upper}$ and higher than $Thresh_{Lower}$ broadcast a new message called *RE-AFF_CH* to re-affiliate nodes belonging to the smaller clusters. Each sensor that receives this message and that belongs to a small cluster should be re-affiliated to the nearest cluster-head based on the received signal strength. Finally, each cluster-head creates a time schedule in which time slots are allocated for intra-cluster communication, data aggregation, inter-cluster communication and maintenance processes. This allows the sensors to remain in sleep state as long as possible and prevents intra-cluster collisions.

4.2 Cluster maintenance

In our contribution, the cluster maintenance process should be triggered in the event of a cluster losing its cluster-head either when the latter exhausts its battery power or migrates towards another cluster. Moreover, the cluster-head's re-election process only concerns clusters that have lost their cluster-head and the future cluster-head would be chosen among the members of the cluster. We adopted this solution so as not to weaken our scheme's performance and to avoid chain reactions which can occur during the launching of the clustering process. Furthermore, the cluster maintenance process is performed in a similar way to the set-up phase, where a random node among the members cluster initiates the clustering process.

5 Evaluation and simulation results

In our experiments, we conducted simulations to evaluate the CES scheme and compare it with LEACH and LEACH-C in terms of the number of nodes alive and the data packets received at the base station during the network lifetime. Simulations have been performed in NS-2 [13] using the MIT_uAMPS ns code extensions [14] to implement the CES scheme. We carried out these simulations with the same scenario presented in [3,4] but with mobile nodes. We considered a network topology with 100 mobile sensors with a sensing range of 25 meters. Sensors are randomly placed in a 100m×100m square area by using a uniform distribution function, and the remote base station is located at position $x = 50$, $y = 175$. At the beginning of the simulation, all the sensors had an equal amount of energy, i.e. the sensors started with 2 Joules of energy. Simulations were carried out until all the sensors exhausted their battery power and the average values were calculated after each round (duration of 20 seconds). After this time, the CES scheme triggered the cluster-head's election process again. Moreover, we performed simulations using two distinct values for threshold $Thresh_{Upper}$: 30, 50, i.e. CES_30 and CES_50, and a fixed value for threshold $Thresh_{Lower}=15$. These values were attributed arbitrarily.

As mentioned above, we used the same energy parameters and radio model as discussed in [3,4], wherein energy consumption is mainly divided into two parts: receiving and transmitting messages. The transmission energy consumption requires additional energy to amplify the signal according to its distance from the destination. Thus, to transmit a k-bit message to a distance d, the radio expends energy as described by the formula (15), where ε_{elec} is the energy consumed for radio electronics, $\varepsilon_{friss\text{-}amp}$ and $\varepsilon_{two\text{-}ray\text{-}amp}$ for an amplifier.   The reception energy consumption is $E_{Rx}=\varepsilon_{elec}\times k$.

$$E_{Tx} = \begin{cases} \varepsilon_{elec} * k + \varepsilon_{friss-amp} * k * d^2 & \text{if } d < d_{Crossover} \\ \\ \varepsilon_{elec} * k + \varepsilon_{two-ray-amp} * k * d^4 & \text{if } d \geq d_{Crossover} \end{cases} \quad (15)$$

Simulated model parameters are set as shown in Table 2. The data size was 500 bytes/message plus a header of 25 bytes. The message size to be transmitted was: k=(500 bytes + 25 bytes)×8= 4 200 bits.

Table 2. Parameters for simulation

Parameter	Value
Network Grid	$(0,0) \times (100,100)$
Base Station	$(50,125)$
ε_{elec}	50 nJ/bit
$\varepsilon_{friss\text{-}amp}$	10 pJ/bit/m^2
$\varepsilon_{two\text{-}ray\text{-}amp}$	0.0013 pJ/bit/m^4
$d_{Crossover}$	87 m
Data packet size	500 bytes
Packet header size	25 bytes
Initial energy per node	2J
Number of nodes (N)	100
Round	20 seconds
Thresh$_{Upper}$	30, 50
Thresh$_{Lower}$	15

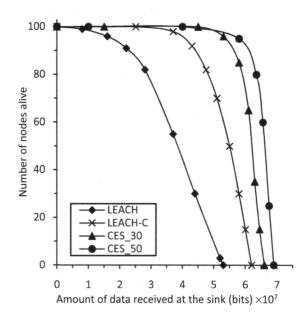

Fig. 3. Number of nodes alive per amount of data received at the sink

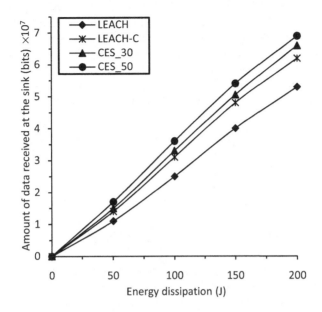

Fig. 4. Amount of data received at the sink according to energy dissipation

Fig.3 shows that CES_30 considerably outperforms LEACH and slightly outperforms LEACH-C in terms of the amount of data sent to the base station during network lifetime, whereas CES_50 largely outperforms them. Moreover, in Fig. 3, the shape of the curves of CES_30 and CES_50 shows that the number of nodes alive degrades rapidly at the end of simulation. That means that the time difference between the demise of the first and last sensor is too small compared to LEACH, where sensors gradually wear out during the network lifetime. On the other hand, Fig.4 illustrates that CES_30 and CES_50 outperform LEACH and LEACH-C in terms of the number of data packets received by the base station with the same total amount of energy.

Our proposed scheme allows the even distribution of consumption among the sensors in the network. Therefore, it maximizes sensor lifetime and minimizes the time difference between the demise of the network's first and last sensor.

6 Conclusion and future work

In this paper, we have proposed a Cluster-based Energy-efficient Scheme (CES) for Mobile Wireless Sensor Networks (MWSNs) which relies on weighing k-density, residual energy and mobility parameters for cluster-head election. The CES scheme carries out a periodical cluster-head election process after each round. Moreover, CES enables the creation of balanced 2-hop clusters whose size ranges between two thresholds: $Thresh_{Upper}$ and $Thresh_{Lower}$.

Simulation results demonstrate that the CES scheme provides better performance than LEACH and LEACH-C in terms of the amount of data received at the base station during the network lifetime, as well as considerably outperforming LEACH in terms of the amount of data sent to the base station with the same amount of energy dissipation.

With these results obtained, the CES scheme can provide good performance for coverage and broadcasting in MWSNs. Therefore, its evaluation could be the subject of future work.

Acknowledgments. The authors would like reviewers for their valuable feedback.

References

1. Chong, C. Y., Kumar, S. P.: Sensor networks: Evolution, opportunities, and challenges. In: Proceedings of the IEEE, vol. 91 N° 8, pp. 1247--1256 (2003).
2. Estrin, D., Govindan, R., Heidemann, J., Kumar, S.: Next century challenges: Scalable coordination in sensor networks. In: *IEEE/ACM* MobiCom'99, pp. 263--270 (1999).
3. Heizelman, W.R., Chandrakasan, A., Balakrishnan, H.: Energy-Efficient Communication Protocol for Wireless Micro Sensor Networks. In: IEEE Proceedings of the Hawaii International Conference on System Sciences (HICSS '00) (2000).
4. Heinzelman, W. R., Chandrakasan, A. P., Balakrishnan, H.: An application-specific protocol architecture for wireless microsensor networks. IEEE Transactions Wireless Communications, vol. 1 N°. 4, pp. 660--670 (2002).
5. Mhatre, V., Rosenberg, C.: Design guidelines for wireless sensor networks: communication, clustering and aggregation. Ad Hoc Network Journal, vol. 2 N°. 1, pp. 45--63 (2004).
6. Ye, M., Li, C., Chen, G., Wu, J.: EECS: an energy efficient cluster scheme in wireless sensor networks. In: IEEE International Workshop on Strategies for Energy Efficiency in Ad Hoc and Sensor Networks (IEEE IWSEEASN2005), Phoenix, Arizona, (2005).
7. Lee, G., Kong, J., Lee, M., Byeon, O.: A Cluster-based Energy Aware Routing Protocol for Sensor Networks. In: Proceedings (466) Parallel and Distributed Computing and Systems, Phoenix, AZ, USA (2005).
8. Muruganathan, S. D., Ma, D.C.F., Bhasin, R.I., Fapojuwo, A. O.: A Centralized Energy-Efficient Routing Protocol for Wireless Sensor Networks. IEEE Communication Magazine, Vol. 43, No. 3, pp. S8--S13 (2005).
9. Lindsey, S., Raghavendra, C.S.: PEGASIS: power efficient gathering in sensor information systems. In: Proceeding of the IEEE Aerospace Conference, Vol. 3, March, pp.1125-1130. Big Sky, Montana (2002).
10. Manjeshwar, A., Agrawal, D. P.: TEEN: A Protocol for Enhanced Efficiency in Wireless Sensor Networks. In: Proceedings of the 1st International Workshop on Parallel and Distributed Computing Issues in Wireless Networks and Mobile Computing, San Francisco, CA (2001).
11. Younis, O., Fahmy, S.: Distributed clustering in ad-hoc sensor networks: A hybrid, energy efficient solution. In: Proceedings of IEEE INFOCOM, Hong Kong (2004).
12. Qing, L., Qingxin, Zhu, Q., Wang, M.: Design of a distributed energy-efficient clustering algorithm for heterogeneous wireless sensor networks. Elsevier Computer Communications 29, pp. 2230--2237 (2006).
13. NS-2, Network Simulator, http://www.isi.edu/nsnam/ns/ns-build.html.
14. MIT_uAMPS LEACH NS2, www.ece.rochester.edu/research/wcng/code/index.html.

Balancing Overhearing Energy and Latency in Wireless Sensor Networks

Byoungyong Lee, Kyungseo Park, Ramez Elmasri

Computer Science & Engineering Department
University of Texas at Arlington
Arlington, TX 76019, USA
{bylee,kpark,elmasri}@uta.edu

Abstract. A WSN (Wireless Sensor Networks) consists of a large number of sensor nodes. Each sensor node has limited battery, small storage, and short radio range. Many researchers have proposed various methods to reduce energy consumption in sensor nodes, since it is difficult to replace sensor node power sources. Generally, a sensor node consumes its energy during processing, receiving, transmitting and overhearing of messages that are directed to other nodes. Among those, overhearing is not necessary for correct operation of sensor networks. In this paper we propose a new synchronized wakeup scheme to reduce the overhearing energy consumption using different wakeup time scheduling for extending sensor network lifetime. The results of our simulation show that there is a trade-off between reducing overhearing energy and delay time. Therefore we propose Double Trees Structure, called DTS, having two routing trees, one based on Short Rings Topology and the other on Long Rings Topology. DTS has multi routing paths from base station to children nodes. If a node which is on the next routing path does not wakeup in time to receive the data, the sender node selects another path to connect to the destination. We can save the wait time until the next destination node wakes up. In the simulation result, our wakeup scheduling reduces overhearing energy consumption more than the S-MAC protocol. Using the double trees structure reduces the delay time.

Keywords: Sensor Network, Wakeup Scheduling, Overhearing Energy.

1 Introduction

Wireless sensor networks are increasingly applied to various physical worlds for surveillance applications. Because large numbers of sensors are typically deployed, the trend has been to decrease the cost of each sensor node. As a result, a sensor node has smaller size than before. Therefore there are various capacity limitations such as the small amount of battery, limited storage, and short radio range [2][4]. Even if each sensor node has small capacity, the large number of sensors can cover a large area by cooperating with each other to form a multi-hop wireless network. Nevertheless low battery power is one of the most crucial problems because it is hard to replace or recharge the battery in each sensor [3]. The life time of a sensor network depends on

Please use the following format when citing this chapter:

Lee, B., Park, K. and Elmasri, R., 2008, in IFIP International Federation for Information Processing, Volume 264; Wireless Sensor and Actor Networks II; Ali Miri; (Boston: Springer), pp. 25–36.

the energy in each senor node. To increase the life time of sensor networks, we need to reduce the energy consumption.

Generally, a sensor node consumes its energy during processing, receiving, transmitting, and overhearing of parts of messages that are not directed to the node. Among those, the energy wasted by overhearing energy consumption is not necessary for correct working of the wireless sensor networks. A characteristic of wireless networks is that some nodes that are not a destination have to receive unnecessary messages because they are within the radio range. This is called overhearing. As node density increases and radio range grows, energy consumed by overhearing also will increase. In order to achieve the purpose of reducing the energy consumption, synchronized wakeup scheduling is used to make a node stay in sleep mode when messages are not directed to the node. We focus on reducing overhearing energy consumption with wakeup scheduling.

For reducing overhearing energy consumption we propose a new wakeup scheme using different wakeup times between neighbor nodes. We call the new wakeup scheme Odd and Even Wakeup Scheduling (OEWS). We compare OEWS with the S-MAC protocol which is one of the popular MAC protocols for sensor networks [1]. In simulation, OEWS shows good results to reduce overhearing energy. This method improves the energy efficiency and increases the sensor network lifetime. OEWS adjusts different wakeup times for sibling nodes. A node in sleep mode will turn off its radio and will not overhear messages. There is a trade-off between energy efficiency and delay time because the node which intends to send the data has to wait until its next destination node wakes up according to a pre-defined synchronized schedule. For reducing delay time, we propose another new tree structure called Double Trees Structure (DTS).

For reducing the data delay time, there are many methods using various wakeup scheduling patterns [14]. However, even if those wakeup patterns are efficient to reduce data delay time, it is hard to adjust those wakeup patterns to OEWS because they didn't consider overhearing energy consumption. Hence, we propose a new routing tree structure called DTS for reducing data delay time on OEWS. DTS has two tree structures called Short Rings Topology (SRT) and Long Rings Topology (LRT). SRT and LRT have the same number of hops from the base station to children nodes. There is no different delay time between SRT and LRT because they have the same number of hops. Therefore, by using multiple paths, we can save the waiting time for children nodes to wake up.

The contribution of this paper is that we explore reducing both overhearing energy and latency together. We propose the OEWS to reduce overhearing energy and DTS for decreasing the latency. Overhearing energy is not necessary for operating the sensor network. Therefore it is important to reduce the overhearing energy for extending the lifetime of sensor networks. In the simulation, OEWS reduces the overhearing energy up to 43% compared to S-MAC protocol. Also, DTS helps OEWS reduce latency up to 30.4% than OEWS without DTS.

The remainder of this paper is organized as follows. In section 2, we provide an overview of our network model. In section 3, we introduce the OEWS for reducing overhearing energy consumption and DTS for reducing delay time. We show the simulation results in section 4. Finally, section 5 presents concluding remarks

2   Network Model

2.1   Radio Model.

For measuring energy consumption in a sensor network, we need an energy model. A sensor node consumes energy by transmitting, receiving and overhearing. E_{tx} is the energy used for transmitting and E_{rx} is the energy used for receiving. We assume the energy model including overhearing energy based on [6]. This radio model calculates the energy spent for one bit to send over a distance 'd' as

$$E_{tx} = E_{txelec} + \epsilon d^2 \qquad (1)$$

$$E_{rx} = E_{rxelec} \qquad (2)$$

E_{txelec}, the energy consumption by transmitter electronics, dissipates *50nJ/bit*. We assume that E_{txelec} is the same as E_{rxelec} (receiving energy) based on [7]. We suppose that ϵ, which is an amplifier characteristic constant, is *100pJ/bit*. The model assumes the radio channel to be symmetric, which means the cost of transmitting a message from A to B is the same as the cost for transmitting from B to A [7]. Overhearing energy consumption is defined by E_{oh} as

$$E_{oh} = E_{rxelec} \qquad (3)$$

In S-MAC protocol [1], when sending data from one sensor node to others, RTS(ready to send), CTS(clear to send), and ACK packets are necessary. Therefore, based on (1), (2), and (3), total energy consumption from node i to node j is represented by the following:

$$Eij = |RTS+DATA| \times (\epsilon di2 + Etxelec) + |CTS+ACK| \times (\epsilon dj2 + Etxelec) + \\ RTS+CTS+DATA+ACK| \times Erxelec + NRTS \times |RTS| \times Erxelec + NCTS \times \\ |CTS| \times Erxelec \qquad (4)$$

Here, d_i is the radio range of node i and d_j is the radio range of node j. N_{RTS} and N_{CTS} is the number of neighbors which overhear the RTS and CTS packets respectively. |RTS,CTS, ACK, or DATA,| is the size of the packet in bits.
In OEWS, total energy consumption is different from S-MAC. OEWS has two kinds of energy models. One is for nodes which have an odd id number and the other is for nodes which have an even id number. Therefore, total energy consumption from node i to node j which both have odd id number (or even id number) is represented by the following:

$$E_{ij} = |RTS+DATA| \times (\epsilon d_i^2 + E_{txelec}) + |CTS+ACK| \times (\epsilon d_j^2 + E_{txelec}) + \\ |RTS+CTS+DATA+ACK| \times E_{rxelec} + N_{odd\_RTS} (or\ N_{even\_RTS}) \times |RTS| \times E_{rxelec} + \\ N_{odd\_CTS} (or\ N_{even\_CTS}) \times | CTS| \times E_{rxelec} \qquad (5)$$

Here, N_{odd_CTS} (N_{even_CTS}) is the number of odd (Even) id neighbors which overhear CTS and N_{odd_RTS}(N_{even_RTS}) is the number of odd (even) id neighbors which overhear RTS.

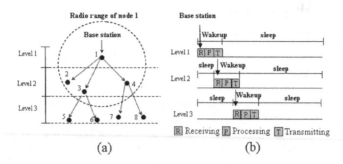

Fig. 1 Example of wakeup model

2.2   Wakeup Model

In wireless sensor networks, we can divide data flow into two directions. In the down direction, data flows from the base station to children nodes. In the up direction, data flows from children nodes to the base station. Our goal is to reduce the overhearing energy consumption when the base station transmits queries or data to children nodes. Hence in our wakeup model, we consider only down directional data flow.

Fig.1 shows our basic wakeup model based on [15]. In Fig.1 (a), the radio range of node 1 covers node 2, node 3 and node 4. Therefore if node 1 intends to send the data to node 2, node 3 and node 4, they could all receive the data from node 1. However, node 5, node 6, node 7 and node 8 could not receive the data from node 1 directly because they are not within its radio range. When node 1 intends to send data to nodes within its radio range, wakeup of nodes not in its radio range such as node 5, 6, 7 and 8 is wasteful of energy, because idle listening consumes energy between 50% and 100% of receiving energy consumption [12]. In [8], Stemm and Katez show that the ratios of idle:receive:send are 1:1.05:1.4 respectively. Also the Digitan 2 Mbps Wireless LAN module specification illustrates idle:receive:send ratios is 1:2:2.5 [9]. Therefore, when node 1 tries to send the data to nodes 2, 3, and 4, the nodes which are not within its radio range such as node 5, 6, 7, and 8 should be synchronized to enter sleep mode for saving idle listening energy. In sleep mode, sensor nodes turn off their power. At the next step, after node 3 or node 4 receives data from node 1, when node 3 or node 4 intend to send the data to nodes within the radio range of sender, the nodes in level 3 wake up and they are ready to receive data from a node in level 2. Nodes in level 1 such as node 1 go to sleep mode again after they send the data to nodes in level 2. All sensor nodes already know their level and their wakeup/sleep synchronized schedule through the setup of the initial tree structure. When the base station makes the initial tree structure, it sets the wakeup duration of each level in advance.

For example, in Fig.1 (b), if the base station decides that nodes wake up for 1 second for data transmission and wakeup for 1 second for data receiving, the nodes in level 1 wake up for 1 second for receiving data from the base station and also wakeup

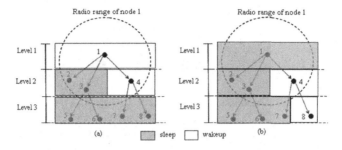

Fig. 2 Odd and even wakeup scheduling example

for another 1 second for transmission to the nodes of level 2. And then nodes of level 1 go to sleep mode. The time interval for data transmission between levels is allocated sufficient time to complete the process, otherwise failure of data transmission will be increased [15].

3   OEWS and DTS

In this section, first we describe the Odd and Even Wakeup Scheduling (OEWS) technique to reduce overhearing energy consumption and then, we describe the Double Tree Structure (DTS) for reducing the delay time caused by OEWS.

3.1   Odd and Even Wakeup Scheduling

We propose the new wakeup schedule named Odd and Even Wakeup Scheduling (OEWS) whose purpose is to reduce the overhearing energy consumption as half the sensor nodes at the same level wake up alternately. In this scheme, sensor nodes having even id number and those having odd id number in the same level have different wakeup time schedule. Sensor nodes having even id number wake up at even time points and sensor nodes having odd id number wake up at odd time. For example, in Fig.2 (a), at a specific even time, sensor node 1 in level 1 can send the data to node 2 and 4 having even number id because node 3 having odd id number is in sleep mode at even time. But at a specific odd time, only node 3 can receive the data from node 1. If node 1 wants to send the data to node 8, node 1 sends the data at the specific even time for sending the data to node 4. Even if node 2 receives the data from node 1 at the same time with node 4, if the destination is not node 2, node 2 goes to the sleep mode. We still save the overhearing energy of node 3 and also we can reduce the wakeup time of node 2. In the next step as shown in Fig.2 (b), after node 4 receives data from node 1, nodes in level 1 fall in sleep mode again and nodes in level 3 wake up for receiving data from nodes in level 2 based on [15]. Nodes 5 and 6 in level 3, they go to sleep mode after they recognize that there is no data from node 3 directed to them. Node 7 and 8 which are children of node 4 wake up alternately in even time and odd time. If node 7 does not receive the data at odd time, node 7 also directly

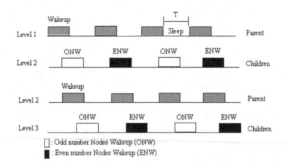

Fig. 3 Odd and even wakeup time scheduling

goes to the sleep mode. And then node 8 can receive the data from node 4 at even time. Hence we can save overhearing energy through the odd-even wakeup schedule. Also, we can decrease the duration of wakeup time for nodes which are not a destination, as they can go to sleep mode if no message is received.

Fig.3 shows the Odd and Even wakeup Scheduling. We use a synchronous wakeup schedule. Level 1 and level 2 in Fig.3 correspond to node 1 and nodes 2, 3, 4 from Fig.2 respectively. First when level 1 transmits data to level 2, the nodes in other levels are in sleep mode. Level 1 follows the schedule for a parent and level 2 follows the schedule for children. A parent node wakes up at every time slot but children nodes wake up alternately. After a node in level 2 receives the data from a node in level 1, the node in level 2 switches from children wakeup time schedule to parent wakeup time schedule. And then the nodes in level 1 are in sleep mode and the nodes in level 3 wakeup and follow the children wakeup time schedule.

In Fig.4, we present the algorithm for a parent node. In lines 1 - 2, a parent node checks its wakeup schedule. If it still follows the children schedule, it changes to parent schedule. In lines 3 – 6, we store the children nodes of the current parent into array *children[j]*. We defined G as an undirected graph and V is a set of sensor nodes. In lines 7 – 11, if parent wakeup time matches with a wakeup time of the target child node, the parent node sends the data to the target child node. Otherwise the parent node waits until its wakeup time matches with the target child wakeup time. In lines 12 – 13, if parent level has time out, parent goes into the sleep mode.

In Fig.4, we also show the algorithm for children nodes. In lines 1 -2 they check their wakeup schedule. If children follow the parent schedule, it changes to children schedule. In line 3, each child tries to find the parent node. In lines 4 – 5, they wait for the data from the parent node. If a specific child node receives the data from the parent node, it becomes a new parent and other sibling nodes in same level go into sleep mode.

3.2   Trade-off between Energy Saving and Delay Time

When we try to send the data from a source to a destination in wireless sensor networks, there is a delay time. We assume the delay time based on [10]. Delay time is the time elapsed between the departure of a data packet from the source sensor and

---

Parent Algorithm

Input : *j=0, n_i* = *number of nodes,*
1: **if** *Schedule* = *children_wakeup_schedule*
2:    **then** *change to parent_wakeup_schedule*
3: **for** $n_i \in V[G]$
4:    **if** $n_i \in$ *children of current parent node*
5:       *children[j]* ← n_i
6:       *j=j+1*
7: **for** target_child_id $\in$ *children[j]*
8:    **if** *target_child_id* ≠ *children[j]*
9:       **then wait( )**
10:    **if** *target_child_id* = *children[j]*
11:       **then send** (*data*)
12: **if** *level_wakeup_time = 0*
13:    **then sleep** (*until next wakeup time*)

Children Algorithm

1: **if** *Schedule* = *parent_wakeup_schedule*
2:    **then** *change to children_wakeup_schedule*
3: **find_parent** ( )
4: **if** parent send the data
5:    **then receive** (data)
6: **if** *level_wakeup_time = 0*
7:    **then sleep** (*until next wakeup time*)

Fig. 4 Algorithm for the parent and children node

its arrival to destination [9]. Therefore we can denoted the delay by $DT(s,d) = (qd+td+pd+wd) \times Nd(s,d)$, where qd is queuing delay, td is transmission delay, pd is propagation delay and wd is waiting delay until the receiver node wakes up. $Nd(s,d)$ denotes total number of data disseminators on the routing path between the source node 's' and the destination node 'd'.

In OEWS, because we use the wakeup scheduling for the nodes to wake up alternately, it causes longer delay time. Therefore we suggest the Double Tree Structure called DTS to reduce the average delay time.

3.3 Double Tree Structure

For OEWS, we make two Rings topologies for the routing tree structure based on [11]. Two Rings topology consists of Short Rings topology called SRT and Long Rings topology called LRT. Rings topology makes a tree structure based on the radio range.

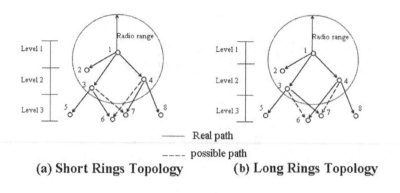

—— Real path

---- possible path

(a) Short Rings Topology   **(b) Long Rings Topology**

Fig. 5 Short rings and long rings topology

(1)   Short Rings Topology and Long Rings Topology

Fig. 5 (a) shows Short Rings Topology (SRT). SRT starts from the base station. In the first step, all nodes within the radio range of the base station become children nodes of the base station. For example, in Fig. 5 (a), the only node within the radio range of the base station is node 1. Therefore node 1 becomes a child node of the base station. In the second step, all nodes within the radio range of node 1 become children nodes of node 1. There are nodes 2, 3, and 4. In the third step, nodes 5, 6 and 7 become children nodes of node 3 because these are within the radio range of node 3. And nodes 6, 7, and 8 become children node of node 4 because these are within the radio range of node 4. Notice that node 6 and node 7 are included as children of both node 3 and node 4. In this case, we can divide them into two Rings topology with distance from parent node to children node. In the view of node 6, node 3 is the closest parent node. Therefore, if node 6 becomes a child of node 3, this topology is Short Rings topology. Otherwise, if node 6 is connected with node 4 which is the most far away from node 6 within the radio range, this topology is Long Rings topology. Fig. 5(b) shows Long Rings Topology.

Our wakeup scheme uses both SRT and LRT in the wireless sensor network. Therefore, it is possible that there are several routing paths. For example, if the base station intends to send the data to node 6, in the first step, the base station sends data to node 1. In the second step, node 1 could send the data to node 3 or node 4. Node 4 can connect to node 6 through the LRT, and node 3 can connect to node 6 through the SRT. Therefore it does not need to wait until odd nodes wake up or even nodes wake up. If there is only one routing path, node 1 has to wait for even nodes wake up or odd nodes wakeup. As a result, we can reduce the delay time.

Even though SRT and LRT are decided by distance, energy consumption for transmitting, receiving and overhearing is the same whether we use SRT or LRT. Because all sensor nodes have the same fixed radio range, radio range of SRT and

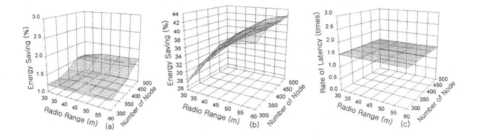

Fig. 6 OEWS vs S-MAC. (a) Energy saving in total energy consumption. (b) Energy saving in overhearing. (c) Latency

LRT is the same. Also, whether we use SRT or LRT, the number of hops from base station to destination is the same. Even if we change the routing path from SRT to LRT or from LRT to SRT, it does not change the number of hops or the radio transmission range.

4   Simulation

In this section, we present simulation results of OEWS and DTS. Our simulation results show that OEWS helps to reduce the overhearing energy consumption and DTS decreases the data delay time. Therefore, we evaluated the efficiency of energy consumption and latency comparing with S-MAC protocol [1]. In S-MAC one of the sources of wasted energy which they tried to reduce is the overhearing energy.

In the experiments, we randomly spread the homogeneous sensors in a $300 \times 300m^2$ sensor field area. All sensor nodes have the same fixed radio range and same energy. We use the DTS with both Short Rings topology and Long Rings topology for the initial routing tree structure. For measuring the energy consumption for transmitting, receiving, and overhearing data, we used the energy model based on [6].

4.1   Efficiency to Reduce Overhearing Energy

In one experiment, we measured the rate of energy saving comparing with S-MAC protocol. Even though S-MAC protocol already reduced the overhearing energy, the experiment result shows that OEWS reduces overhearing energy more than S-MAC. Fig.6 (a) and (b) show the energy saving results of OEWS comparing with S-MAC. We then increased the number of sensor nodes from 300 to 600 and the radio range from 30m to 60m. In Fig.6 (a), we compare the energy saving rate in total energy consumption including transmitting, receiving and overhearing. This result shows that OEWS reduces up to 1.7% more energy than S-MAC protocol. With the high density of sensor nodes, OEWS produces more saving of overhearing energy consumption. Fig.6 (b) shows the result where only compare the overhearing energy

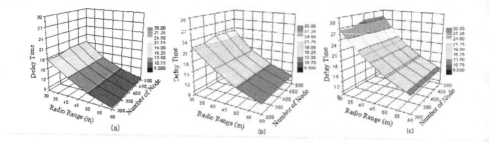

Fig. 7 (a) Delay time on S-MAC. (b) Delay time on DTS. (c) Delay time on OEWS

with S-MAC protocol. We see that OEWS can save the overhearing energy up to 43% more than S-MAC protocol. In this result, the more density of sensor nodes, the larger the decrease in overhearing energy consumption. Therefore, OEWS is more suitable for high density sensor networks than low density sensor networks.

4.2   Effect on Latency between OEWS and S-MAC

In this section, we analyze the data latency between OEWS and S-MAC protocol. In Fig.3, for example, when a node in level 1 detects some events happening between the T period which is the duration of sleep, the node in level 1 waits until its next wakeup time. The probability of occurring event between T periods is uniformly distributed. Therefore we represent the uniform distribution between A and B as $X \sim U[A,B]$ based on [14]. X is random delay time. A and B are the smallest delay time and the largest delay time respectively. S-MAC has very similar performance to wakeup synchronized time where all the nodes in the sensor networks wake up and go to sleep mode at the same time with the same wakeup time. Hence delay time of S-MAC is represented by the following:

$$X \sim U[\,(h-1)T,\, hT\,]$$ (6)

Therefore, average delay time is :

$$E(X) = (h - \tfrac{1}{2})T$$ (7)

In formulas (6) and (7), h means the number of hops.

In OEWS, half of nodes of all sensor nodes wake up and then the other half wake up. Therefore a node in OEWS takes two times waiting time until the next wakeup time. Therefore, we can represent OEWS delay by the following:

$$X \sim U[\,(h-1)T,\, 2hT\,]$$ (8)

The average delay time of OEWS is the following:

$$E(X) = (\tfrac{3}{2}h - \tfrac{1}{2})T$$ (9)

Fig.6 (c) shows the result of a average latency in OEWS comparing with S-MAC. We simulated with number of nodes from 300 to 600 and radio range from 30 to 60m.

In this environment, latency of OEWS is 1.51 times latency of S-MAC protocol. Hence this simulation shows that there is a trade-off between overhearing energy consumption and latency. In the next section, we show that DTS which we proposed as a routing tree structure reduces the latency.

4.3   Effect on DTS in OEWS

We use the new routing tree structure called Double Tree Structures (DTS) to reduce the latency on OEWS. In this section, we compare latency between OEWS with DTS and OEWS without DTS. Fig. 7(b) shows the delay time of OEWS with DTS and Fig. 7(c) shows the delay time of OEWS without DTS. From these results, we know that DTS works to reduce the latency. DTS has Long Rings topology and Short Rings Topology. Therefore, there are two routing trees in the sensor network. If some nodes have alternative paths to the destination, we can reduce the waiting time. Hence average delay time of DTS with OEWS is represented by the following:

$$E(X) = \frac{(h - \frac{1}{2})T + (\frac{3}{2}h - \frac{1}{2})T}{2} \tag{10}$$

where, h is number of hops and T is the time duration of sleep mode.

Also, we compare latency between OEWS with DTS and S-MAC. When we compare S-MAC delay time on Fig. 7(a) with OEWS without DTS on Fig. 7(c), even though OEWS reduce overhearing energy, OEWS has more delay time by about 50%. But when we used DTS on OEWS, OEWS with DTS has more delay time over S-MAC by about 13%. Therefore, we improved the trade-off between overhearing energy consumption and latency with proposed OEWS and DTS.

5   Conclusion

In this paper, we have proposed OEWS for reducing the overhearing energy consumption with different wakeup times and DTS for decreasing the latency of OEWS with double tree structure. Even if OEWS advantage is to reduce the overhearing energy consumption, there is a delay time because of a trade-of between energy saving and delay time. But DTS is useful to reduce the delay time. In DTS, Long Rings topology and Short Rings topology have the same number of hops from any node to the base station. Therefore using either Long Rings topology or Short Rings topology for routing path, the number of hop does not have effects on delay time.

Our simulation results also show OEWS and DTS have good performance. OEWS and DTS are more suitable for high density sensor network. Overhearing energy consumption is high when nodes are having many neighborhood sensor nodes.

References

1. W. Ye, J. Heidemann, D. Estrin, "An energy-efficient MAC protocol for wireless sensor netowkrs"in INFOCOM 2002: Proceedings of the Twenty-First Annual Joint Conference of the IEEE Computer and Communications Societies, Vol. 3, IEEE, June 2002   .
2. Mohamed A. Sharaf, Jonathan Beaver, Alexandros Labrinidis, Panos K. Chrysanthis, "Balancing energy efficiency and quality of aggregate data in sensor networks" VLDB journal, 2004 13:384-403.
3. Ramanan Subramanian, Faramarz Fekri, "Sleep scheduling and lifetime maximization in sensor networks: fundamental limits and optimal solutions" Proceedings of the fifth international conference on Information processing in sensor networks, IPSN 06.
4. Ossama Younis, Sonia Fahmy, "Distributed clustering in ad-hoc sensor networks:a hybrid, energy-efficient approach," IEEE Transaction on Mobile Computing Vol.3, No.4, 2004.
5. Byoungyong Lee, Kyungseo Park, Ramez Elmasri, "Energy Balanced In-Network Aggregation Using Multiple Trees in Wireless Sensor Networks", Consumer Communications and Networking Conference, Jan. 2007.
6. Prithwish Basu, Jason Redi, "Effect of overhearing transmissions on energy efficiency in dense sensor networks", Proceedings of the third international symposium on Information processing in sensor networks, IPSN 04.
7. W.R. Heinzelman, A. Chandrakasan, H.Balakrishnan, "Energy-efficient communication protocol for wireless microsensor networks", Proceedings of the 33rd Hawaii International Conference on System Sciences 2000.
8. Mark Stemm, Randy H Katz, "Measuring and reducing energy consumption of network interfaces in hand-held devices," IEICE Transactions on Communications, vol. E80-B, no.8, pp1125-1131, Aug.1997.
9. Oliver Kasten, Energy Consumption, http://www.inf.ethz.ch/~kasten/ research/bathtub/ energy_consumption.html, Eldgenossische Technische Hochschule Zurich.
10. Habib M. Ammari and Sajal K. Das, "Trade-off between energy savings and source-to-sink delay in data dissemination for wireless sensor networks," Proceedings of the 8th ACM international symposium on Modeling, analysis and simulation of wireless and mobile systems, MSWiM '05.
11. S. Nath, P. Gibbons, S. Seshan, Z. Anderson, "Synopsis Diffusion for Robust Aggregation in Sensor Networks," In Proc. 2nd ACM SenSys, pp 250-262,2004.
12. Qin Wang, Zygmunt J. Haas, "BASS: an adaptive sleeping scheme for wireless sensor network with bursty arrival," Proceedings of the 2006 international conference on wireless communications and mobile computing, IWCMC '06.
13. Wei Ye, John Heidemann, Deborah Estrin, "Medium access control with coordinated adaptive sleeping for wireless sensor networks," IEEE/ACM Transaction on Networking, vol. 12, No.3, June 2004.
14. A. Keshavarzian, H. Lee, L. Venkatranman, "Wakeup Scheduling in Wireless Sensor Networks," MobiHoc '06, 2006, Florence, Italy.
15. S. Madden, M. Franklin, J. Hellerstein, W. Hong "TAG: a Tiny Aggregation Service for Ad-Hoc Sensor Networks," In OSDI, 2002.

Transmission Power Management for IR-UWB WSN Based on Node Population Density

Fernando Ramírez-Mireles

Instituto Tecnológico Autónomo de México (ITAM)
Río Hondo 1, Col. Tizapán San Angel,
México City, D.F. C.P. 01000,
México
ramirezm@ieee.org
http: // www.geocities.com / f _ ramirez _ mireles

Abstract. We propose a method to manage transmission power in nodes belonging to a wireless sensor network (WSN). The scenario contemplates uncoordinated communications using impulse radio ultra wideband (IR-UWB). Transmission power is controlled according to the statistical nature of the multiple access interference (MAI) produced by the nodes in the close vicinity of the communicating nodes. The statistical nature of the MAI is a function of the node population density within the area of coverage of the WSN. We show that when the node population density is high enough transmission power savings are possible.

Keywords: UWB, impulse radio, ad hoc networks, sensor networks.

1 Introduction

Wireless ad hoc networks (WAHN) are flexible networks for which there is no need of a central coordinator and for which the numbers of nodes and the topology of the network are not predetermined. A WSN is a type of WAHN composed of nodes with sensing capability. There are important differences between WSN and WAHN [1]. WSAN have a larger number of nodes and are deployed in close proximity to the phenomena under study, the nodes mainly use a broadcast communication paradigm and the topology of the network can change constantly due, for example, to nodes prone to fail (usully nodes have limited power, computational capabilities and memory).

The UWB is an indoor communication technique currently under intense research activity [2] due to many attributes,[1] including its robustness against multipath conditions, its high capacity in a multiple access environment, the capability to achieve high transmission rates using a low amount of power, and, for pulse based UWB, the possibility of operating using a carrier-less modulation.

---

[1] According to [3] a signal is considered of UWB nature if the 10 dB bandwidth of the signal is at least 20% of its center frequency, or if this 10 dB bandwidth is at least 500 MHz.

Please use the following format when citing this chapter:

Ramírez-Mireles, F., 2008, in IFIP International Federation for Information Processing, Volume 264; Wireless Sensor and Actor Networks II; Ali Miri; (Boston: Springer), pp. 37–48.

In particular, IR UWB [4] uses communications signals composed of ultra short pulses, and multiple access is achieved providing nodes with different spreading codes. The IR-UWB has been proposed for WAHN [5]. The many desirable characteristics of IR-UWB can be used in WAHN and WSN for simultaneous communication, ranging and positioning [6].

In this work we propose a method to manage the transmission power of a transmitting node. Communication signals use binary pulse position modulated (PPM) for data modulation and time hopping (TH) for code modulation. Spread spectrum multiple access (SSMA) is achieved using different TH codes for different users. We consider a WSN where the nodes know the spreading codes of their close neighbors, and the nodes communicate with the other nodes broadcasting messages. There can be some coordination between the nodes, e.g., they know the spreading codes of their close neighbors, and they can estimate roughly how many nodes are within their own neighborhood, but transmission are uncoordinated, e.g. with asynchronous transmission times and no power control.

Multiple nodes broadcasting their messages will generate MAI to other nodes, and this MAI will degrade the performance of the communication links. To preserve battery life, transmission power is controlled according to the statistical nature of the MAI produced by the nodes in the close vicinity of the communicating nodes. The statistical nature of the MAI is a function of the node population density within the area of operation of the WSN. We show that when the node population density is high enough transmission power savings are possible.

2 MAI statistical nature and its impact on performance

Consider a system with $k = 1, 2, 3, \ldots, N_u$ nodes, and that nodes $2, 3, \ldots, N_u$ are simultaneously broadcasting to other nodes. In particular, suppose that node 2 is attempting to transmit to node 1, while the others $N_u - 2$ nodes are broadcasting in the vicinity of node 1. Lets denote the received signal power at node 1 as $(Q_{k,1})^2 P_k$, $k = 2, 4, \ldots, N_u$, where P_k is the transmitted power of node k and $(Q_{k,1})^2$ is a factor (that can be random) reflecting the attenuation from transmitter k to receiver 1. If we ignore the effects of the noise, the signal-to-interference-ratio (SIR) at the *input* of node one's correlation receiver is

$$\text{SIR}_{\text{in}}(N_u) = \frac{(Q_{2,1})^2 P_2}{\sum_{k=3}^{k=N_u}(Q_{k,1})^2 P_k}. \tag{1}$$

The statistical nature of the MAI at the *output* of the correlation receiver depends on the numbers of nodes $N_u - 2$ producing it and also on the signal structure. In our case we consider binary TH-PPM signals[2]

$$S^{(k)}(t) = \sum_{m=0}^{N_s-1} p(t - mT_f - c_m^{(k)}T_c - T_d d_{\lfloor m/N_s \rfloor}^{(k)}), \tag{2}$$

[2] This signal set and its performance under different scenarios are studied in detail in [7] [8].

where T_f is the frame time, T_c is the code time shift, and T_d is the data time shift. The $c_m^{(k)}$ is the pseudorandom time-hopping sequence for user k with a range $0 \leq c_m^{(k)} < N_h$ and with sequence period N_p. The $d_{\lfloor m/N_s \rfloor}^{(k)}$ is the data of user k that can be 0 or 1 and changes every N_s hops (see fig. 1). Note that the data symbol changes every N_s frames, therefore the symbol duration is $N_s T_f$ and the bit rate is $R = \frac{1}{N_s T_f}$.

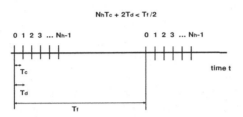

Fig. 1. Relation of TH-PPM parameters.

We consider two forms of pulse shape, with pulse parameters shown in table 1. The first is the second derivative of a Gaussian pulse

$$p(t) = \left[1 - 4\pi \left[\frac{t}{t_n} \right]^2 \right] \exp \left(-2\pi \left[\frac{t}{t_n} \right]^2 \right), \tag{3}$$

for $-T_w/2 \leq t \leq T_w/2$, where $t_n = 0.2877$ ns is a parameter that determine the pulse duration $T_w \simeq 0.7$ ns. The second pulse shape is based on a gated sine wave

$$p(t) = \sin \left(2\pi \frac{Q}{T_w} t \right), \tag{4}$$

for $-T_w/2 \leq t \leq T_w/2$, $T_w = 2.0$ ns, where $Q = 10$ is a positive integer, resulting in a signal spectrum centered at $\frac{Q}{T_w} = 5$ GHz.

For the signals in (2) the SIR at the *output* of the correlator can be written [7] [8]

$$\mathrm{SIR}_{\mathrm{out}}(N_u) = \frac{(Q_{2,1})^2 P_2 A_{2,1}}{\sum_{k=3}^{k=N_u} (Q_{k,1})^2 P_k C_{k,1}}. \tag{5}$$

where $A_{2,1}$ and $C_{k,1}$ are normalized autocorrelation and cross correlation factors. The $A_{2,1}$ is produced by the desired signal and the $C_{k,1}$ by the interference.

Given a pulse shape, the parameters mainly determining the MAI statistical nature are the number of pulses per symbol N_s, and the frame time between pulse transmissions T_f. It has been verified that under certain conditions (e.g. if N_s is large enough), an increase of N_u will cause the statistical nature of the

Table 1. Values for TH-PPM time parameters.

Parameters	Gaussian Pulse	Gated sine wave
T_w	0.7 ns	2.0 ns
T_f	70, 150, 250 ns	200 ns
T_d	0.156 ns	0.0995 ns
T_c	0.9 ns	0.1 ns
N_h	8	200

MAI to tend to Gaussianity [9]. However, when those conditions are not met, the MAI will have a probability density function that in general has thicker tails than the Gaussian distribution [10], producing a higher bit error rate (BER) than the Gaussian MAI, for the same SIR value.

Fig. 2 shows the difference in performance between a MAI that is not Gaussian (proposal approximation (PA)) and a MAI that is Gaussian (Gaussian approximation (GA)). Notice that as N_u is increased the BER is degraded since the SIR in (5) decreases. Also notice that as N_u is increased the PA tends to behave like the GA [9] [11] due to the central limit theorem in probability [12]. Fig. 2 verifies that BER for PA is higher than for GA, and shows the SIR gap between PA and GA.

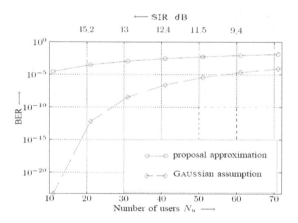

Fig. 2. BER for $T_f = 100$ ns, $N_s = 4$, and $10 \leq N_u \leq 70$ (elaborated in part with data taken from [11]).

3   Using the MAI statistical nature to control transmission power

Each component of the MAI at the output of the correlation receiver is a random variable (r.v.) that depends on other r.v.'s such as the transmissions delays, the spreading codes, the users data, and the channel statistics. The MAI can be written as [7] [8]

$$\alpha = \sum_{k=3}^{N_u} \alpha^{(k)}$$

$$\alpha^{(k)} \triangleq \sum_{k=3}^{N_u} \sum_{m=0}^{N_s-1} A^{(k)}[R(\lambda_m^{(k)}) - R(\lambda_m^{(k)} - T_d)], \qquad (6)$$

where $\lambda_m^{(k)} = [c_{m-\Phi}^{(k)} - c_m^{(1)}]T_c + d^{(k)}T_d + \tau_k$ is a weighted sum of uniformly distributed r.v.s $c_j^{(k)}$, $d_{\lfloor m/N_s \rfloor}^{(k)}$ and τ_k, $R(\cdot)$ are cross correlation terms, τ_k are random transmission delays, and Φ is a r.v. that depends on the transmission delays. We notice that the r.v.'s $[c_{m-\Phi}^{(k)} - c_m^{(1)}]$ for distinct values of m are conditionally independent, given the value of the time shift parameter Φ. We also notice that the r.v.'s $\alpha^{(k)}$ for distinct values of k are conditionally independent, given the code $\{c_m^{(1)}\}$ of user one.

Given the signal structure in (2) and the channel conditions, the statistical nature of the MAI in (6) is a function of the node population density within the area of coverage of the WSN.

We now use the data in fig. 2 to illustrate how transmission power savings are possible when the node population density is high enough. Let's look at the results for SIR=12.4 dB. If MAI satisfy the GA the BER is about 10^{-7}, but if MAI satisfy the PA then BER is about 10^{-3}. Let's say we have a target BER of 10^{-3}. If we are operating in the so called PA regime, a SIR=12.4 dB is just enough, but if we are operating in the so called GA regime, a lower SIR would be fine.

This means transmitter node 2 could reduce its power transmission and still satisfy the target BER. Hence, if node 2 somehow knows that the GA regime applies, it will proceed to reduce its transmission power. To determine if the GA regime can be applied or not, it would need to know how many nodes are in its own vicinity, i.e., it would need to know the node population density in its neighborhood.

The GA regime is reasonable for SSMA systems with both low per-user data rate and a large number of users. Since WSN transmit at low bit rate and are usually densely populated, transmission power could be controlled according the statistical nature of the MAI produced by the nodes in the close vicinity of the target node.

We now proceed to investigate what would be the critical number of nodes in the vicinity from which the GA regime can be applied.

4 Determining the number of nodes to apply the GA regime

To study how the domain of validity of the GA regime changes we consider 3 scenarios: An ideal propagation channel with perfect and imperfect power control,[3] as well as a multipath channel with "perfect average" power control. In our method we fix certain signal design parameters (e.g. pulse shape and duration, pulse position and frame time T_f), and find for which pairs (N_s,N_u) the Gaussian assumption can be considered valid using an entropy test to determine Gaussianity.

4.1 Gaussianity Test.

In this section we describe a procedure to determine the Gaussianity of a sample α of MAI. More specifically, we collect a sample α in (6) and calculate an entropy function which is then compared with the entropy of a Gaussian r.v.

Let's consider the following hypothesis testing

$$\mathcal{H}_0 : \alpha \sim N(0,\ \sigma^2)$$
$$\mathcal{H}_1 : \alpha \sim \text{ not } N(0,\ \sigma^2)\,, \tag{7}$$

where α is assumed to have zero-mean and variance σ^2, and $N(\cdot)$ is the normal distribution. $\mathcal{H}_1$ say's that the distribution of α is not Gaussian. By exploiting the fundamental fact that a r.v. has maximum differential entropy if and only if its Gaussian distributed [13], the following equivalent hypothesis testing problem can be established:

$$\mathcal{H}_0 : \text{entropy}(\alpha) = \tfrac{1}{2}\ln\left(2\pi e\sigma^2\right)$$
$$\mathcal{H}_1 : \text{entropy}(\alpha) < \tfrac{1}{2}\ln\left(2\pi e\sigma^2\right) \tag{8}$$

where e is the base of $\ln(\cdot)$.

An elaboration of hypothesis testing in (8) by [14] gives

$$\mathcal{H}_0 : K_{nm} \ \to \ \sqrt{2\pi e} = 4.1327\ldots$$
$$\mathcal{H}_1 : K_{nm} \ < \ \sqrt{2\pi e} \tag{9}$$

where

$$K_{nm} = \frac{n}{2m\overline{\sigma}}\left\{\prod_{i=1}^{n}\left(\alpha_{(i+m)} - \alpha_{(i-m)}\right)\right\}^{1/n} \tag{10}$$

is a normalized *Gaussianity index*, and where $\overline{\sigma}^2 = \frac{1}{n}\sum_{i=1}^{n}\left(\alpha_i - \overline{\alpha}\right)^2 \neq 0$, is the sample variance.

---

[3] Results in fig. 2 are for ideal propagation conditions with perfect power control.

To establish the Gaussianity test a critical value K^* is proposed so that when $K_{nm} \geq K^*$, with certain probability, we accept that α is a Gaussian r.v. For this purpose we use $K^* = 3.35$ that has a 95% confidence level for $n = 50$ and $m = 5$, and was originally proposed in [14]. With these assumptions we can rewrite the final hypothesis testing problem as

$$\mathcal{H}_0 : K_{nm} \geq K^*$$
$$\mathcal{H}_1 : K_{nm} < K^*. \tag{11}$$

For a given $p(t)$ and T_f, we can use this test to define the *Gaussian regime* as the set of all pair of points (N_s, N_u) such that the normality index $K_{n,m} \geq K^*$.

5   Simulation results.

5.1   scenarios

Fig. 3 depicts the power profiles for the three scenarios considered: 1) Ideal propagation with perfect power control, where the power of each component in the MAI is equal to a constant, 2) Ideal propagation with imperfect power control, where the power of each component $k = 3, 4, \ldots, N_u$ in the MAI is a r.v., and 3) A multipath channel with "perfect average" power control, where the power of each node $k = 2, 3, 4, \ldots, N_u$ is a r.v. For the random powers we consider multipath channel models [15] [16] with two power profiles: line-of-sight (LOS) and not-line-of-sight (NLOS). We use transmitter-receiver distances $D = 3, 6, 9$ m for LOS and $D = 1, 2, 3$ m for NLOS. We produce 49 profiles for each distance and average over the 3 distances in each case.

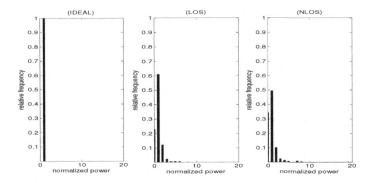

Fig. 3. Power profiles considered in this work.

5.2   Calculation of normality index

Samples of the r.v. α were generated using Matlab. The α in (6) can be seen as a double sum of i.i.d r.v.'s containing terms $R(\cdot)$.

Next we evaluate the normality index $K_{n,m}$ to obtain the region where α is considered a Gaussian r.v., i.e., $K_{50,5} \geq 3.35$. The boundary of the Gaussianity region is determined by those (N_s, N_u) where the entropy test is satisfied.

We simulate several cases considering a large amount of permutations of (N_s, N_u). Since the entropy estimator is itself a r.v., the random boundary is actually a region determined by many random realizations. To study the boundary with a fixed T_f value we generate 100 of such realizations. To study boundary changes for different T_f we smooth the plots by averaging over 10 of such realizations.

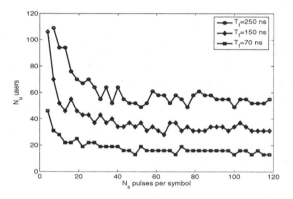

Fig. 4. Boundaries of the Gaussianity region for different T_f with perfect power control (Gaussianity region is on top and to the left of the boundary).

5.3   Ideal propagation with perfect power control

Fig. 4 shows how the boundary changes for different T_f using the Gaussian pulse in (3). Comparing our results with previous works, we found a good match. For example, fig. 2(a) in [11] shows that for $N_s = 8$ and $T_f = 150ns$ about $N_u = 50$ users are needed to be in the Gaussian region, a number that agrees with results in fig. 4. As another example, propositions 6 and 7 in [9] shows that for a fixed N_s the larger T_f is the larger N_u should be to reach Gaussianity, a situation that agrees with results in fig. 4.[4] Intuitively, a shorter T_f results in more pulse

---

[4] Notice that [9] uses a signal format and parameters in which N_s, N_h and T_f are coupled, i.e., they define $T_f = N_h T_c$ and a processing gain $N = N_s N_h$, hence $T_f = N T_c / N_s$. By keeping N fixed, a low value of N_s implies a large value of T_f. Notice that in our case we consider PPM with $T_f/2 > N_h T_c + 2T_d$, hence N_h and N_s can be changed without necessarily affecting T_f (see fig. 1).

collisions than a longer T_f, hence with a shorter T_f Gaussianity is achieved with lower values of N_s and/or N_u.

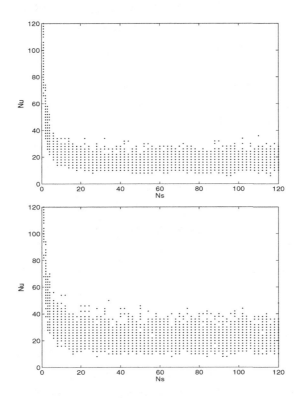

Fig. 5. Boundaries of the Gaussianity region with imperfect power control. Top: LOS power profile. Bottom: NLOS power profile.

5.4   Ideal propagation with imperfect power control

Fig. 5 shows the range of pairs (N_s, N_u) defining the random boundary using the pulsed sinusoid. We notice that there is an asymmetry on the range of values for N_s and N_u needed to reach Gaussianity. On one hand, it is observed that for low N_s we need a large N_u to reach Gaussianity. This can be explained by recalling that $\alpha^{(k)}$ for $k = 3, 4, \ldots, N_u$ are dependant r.v.'s trough user one's code $\{c_m^{(1)}\}$. On the other hand, it is observed that N_u reaches a sort of minimum floor as N_s is increased. This can be explained recalling that $\lambda_m^{(k)}$ for $m = 0, 2, \ldots, N_s - 1$ are dependant r.v.'s trough Φ, and that τ_k, being uniformly distributed over $\left[-\frac{T_f}{2}, \frac{T_f}{2}\right]$, is the largest component in $\lambda_m^{(k)}$ compared to $[c_{m-\Phi}^{(k)} - c_m^{(1)}]T_c + d^{(k)}T_d$.

Results for this case also show that the boundary with imperfect power control have a higher spread of (N_s, N_u) values than the boundaries with perfect power control. This is expected since the weak components of the MAI are masked by the strong ones, hence it takes more r.v. to get to the Gaussianity region. We also notice that boundaries for NLOS have a higher spread than the boundaries for LOS. This is explained observing that the NLOS power profile in fig. 3 have a higher spread than the LOS power profile.

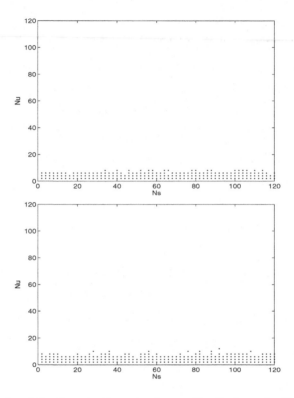

Fig. 6. Boundaries of the Gaussianity region in a multipath channel with average power control. Top for LOS and bottom for NLOS.

5.5   Multipath channel with perfect average power control

Fig. 6 shows the random boundary using the pulsed sinusoid in (4). We notice that in this case there is also an asymmetry on the range of values for N_s and N_u needed to reach Gaussianity, i.e., the range of N_u needed to reach Gaussianity is approximately constant for all values of N_s considered, and this range contains just a few interferer users, hence, in a dense multipath Gaussianity can be reached with few values of both N_s and N_u. These results are consistent with previous studies where a single-user UWB signal in multipath can be modeled as a non-stationary Gaussian random process [17].[5]

6 Discussion and conclusions.

In this paper we propose a method to manage transmission power in nodes belonging to an IR-UWB WSN.

Transmission power is controlled according to the statistical nature of the MAI produced by interfering nodes in the close vicinity of the target node. The statistical nature of the MAI is a function of the node population density within the area of coverage of the WSN.

We established a statistical test to determine when the MAI at the output of a correlator can be considered a r.v. with a Gaussian distribution. Using an entropy point estimator we determine the minimum number of users N_u and number of pulses per symbol N_s necessary to consider the MAI component as a Gaussian r.v. We show that when the node population density is high enough MAI can be considered as a Gaussian random variable and in that case transmission power savings are possible. Our analysis assumes operation at low to medium SNR values.

One issue (noted by a reviewer) is the mechanism that allows a node to know the spreading code of the other nodes and to know the population density in its vicinity. This power management technique can be used during periods of stability in the network, e.g., once the topology and node population are relatively stable, and after the network initial communication set up has been completed. For the nodes to remain simple, some form of centralized control and monitoring should be used, but further research is needed to find specific protocols that can handle operation in relatively stable conditions as well as during transitions. One potential advantage of our technique relatively to other work in the literature [18] [19] is that we do not require to estimate the power of the interference but just to register the number of the interferers.

Another issue (noted by a reviewer) is whether or not the use of the proposed power-control technique could lead to 'oscillatory' behavior in which the MAI statistics switches back and for from the GA regime to the PA regime as nodes increase and decrease the power levels. Results in fig. 5 for ideal propagation with imperfect power control implicitly consider this situation,[6] showing that the GA regime is stable for Nu high enough (e.g. $N_u \geq 60$ nodes in fig. 5). However, this value of N_u can change with the statistics of the power profile.

---

[5] Notice, however, that fig. 7 in [9] shows that a higher N_u is needed to reach Gaussianity. But results in [9] are for a multipath channel with a limited number of discrete paths, while in this work we are using a dense multipath channel with a continuous impulse response.

[6] In fig. 5 the received power from the different nodes fluctuates from low to high to low again according to the power profile in fig. 3.

References

1. I. F. Akyildiz, W. Su, Y. Sankarasubramaniam, and E. Cayirci, A survey on sensor networks, in IEEE Communications Magazine, Vol. 40 (2002), pp. 102- 114.
2. Various authors, Special issue on ultra-wideband communications, in IEEE J. Select. Areas Commun., vol. 24 (2006).
3. U.S. Federal Communications Commission, First Report and Order for UWB Technology, U.S. Federal Communications Commission, (2002).
4. R. A. Sholtz, Multiple-access with time-hopping impulse modulation, in Proc. Military Communications Conf. (1993) pp. 447-450.
5. L. De Nardis, P. Baldi, and M.-G. Di Benedetto, UWB Ad-Hoc networks, in IEEE Conference on Ultra Wideband Systems and Technologies, (2002), pp. 219 - 223.
6. L. De Nardis, and M.-G. Di Benedetto, Joint communication, ranging, and positioning in low data-rate UWB networks, in 2nd. Workshop on positioning, navigation and communication and 1st ultra-wideband expert talk, (2005), pp. 191-200.
7. F. Ramírez-Mireles, Performance of ultrawideband SSMA using time hopping and M-ary PPM, in IEEE J. Select. Areas Commun., vol. 19 (2001), pp. 1186-1196.
8. F. Ramírez-Mireles, Performance of UWB SSMA using orthogonal PPM-TH over dense multipath channels, in Springer's Telecommunications Systems, vol. 36 (2007), pp. 107-115.
9. J. Fiorina and W. Hachem, On the asymptotic distribution of the correlation receiver output for time-hopped UWB signals, in IEEE Trans. on Signal Processing, vol. 54 (2006), pp. 2529 - 2545.
10. N. C. Beaulieu and S. Niranjayan, New UWB Receiver Designs Based on a Gaussian-Laplacian Noise-Plus-MAI Model, in IEEE International Conference on Communications, (2007), pp. 4128 - 4133.
11. Y. Dhibi and T. Kaiser, On the impulsiveness of multiuser interferences in TH-PPM-UWB systems, in *IEEE Trans. on Signal Processing*, vol. 54 (2006), pp. 2853 - 2857.
12. A. Papoulis, Probability, random variables, and stochastic processes, New York:. McGraw Hill (1965).
13. Robert B. Ash, Information theory, Dover Publications, New York, (1990).
14. O. Vasicek, A test for normality based on sample entropy, Journal of the Royal Statistical Society Series B, vol. 38 (1976), pp. 54-59, 1976.
15. W. Turin, R. Jana, S. Ghassemzadeh, C. Rice, and V. Tarokh, Autoregressive modeling of an indoor UWB channel, in Proc. UWBST Conf.,(2002), pp. 71-74.
16. S. Ghassemzadeh, R. Jana, C. Rice, W. Turin, and V. Tarokh, A statistical path loss model for in-home UWB channel, in *Proc. UWBST Conf.*, (2002), pp. 59-64.
17. Q. T. Zhang and S. H. Song, "Parsimonious Correlated Nonstationary Models for Real Baseband UWB Data", in *IEEE Trans. on Vehic. Technol.*, vol. 54, pp. 447-455, Mar. 2005.
18. F. Cuomo, C. Martello, S. Baiocchi, F. Capriotti, "Radio Resource Sharing for Ad Hoc Networking with UWB", in *IEEE JSAC*, vol. 20 (2002), pp. 1722-1732.
19. H. Jiang, W. Zhuang, and X. Shen, "Distributed medium access control for next generation CDMA wireless networks", in *IEEE Wireless Communications,* vol. 14 (2007), pp. 25-31.

Power-On Controller for high lifetime wireless sensor nodes

Mickaël Cartron, Nathaniel Seymour and Yannick Bonhomme

CEA, LIST, Laboratoire de Fiabilisation des Systèmes Embarqués, Boîte Courrier 94, Gif-sur-Yvette, F-91191 France
mickael.cartron@cea.fr

Abstract. Power savings are nowadays crucial in embedded system contexts such as Wireless Sensor Networks (WSN) in order to increase the lifetime of sensor nodes. In this paper, we propose a new hardware structure called "Power-On Controller" (POC) for applying advanced control strategies for the "idle to active" node state transition. The proposed POC allows an optimization of power control by using event accumulation and spatial selectivity mechanisms. These new features allow to reduce the dynamic power consumption of roughly 60% compared to state-of-the-art power management solutions for a typical WSN applicative context, without altering the quality of service. The POC structure can be easily integrated in any sensor node based on system-on-chip design.

Introduction

Nowadays, distributed wireless devices are deployed more and more for multiple high value-added applications [19] (e.g. metering, monitoring, surveillance or tracking) in different contexts (e.g. industrial, military or medical). This success is directly linked to their low cost (i.e. for multi-purpose hardware platform, for maintenance and for network deployment and placement). Indeed, wireless nodes are self-powered with self-organized communication capability and as a consequence, they do not require any fixed infrastructure, neither for power, nor for communication.

Another specificity of wireless sensor networks is their applications which request low processing activity ("control-oriented applications") with no background tasks or intensive calculation.

Besides that, the lifetime of sensor networks nodes should be as high as possible under given design constraints (e.g. volume or weight) due to the environment of interest. This assertion leads to the fact that these systems are highly constrained in terms of energy consumption. In order to enhance the system autonomy, much progress has been obtained in battery technology (i.e. over 400% gain in 20 years [5]), in physical integration [25] and in dynamic and static power optimizations.

The work reported here was performed as part of the ongoing research program µSWN FP6-2005 IST-034642 and funded by the European Social Fund (ESF).

Please use the following format when citing this chapter:

Cartron, M., Seymour, N. and Bonhomme, Y., 2008, in IFIP International Federation for Information Processing, Volume 264; Wireless Sensor and Actor Networks II; Ali Miri; (Boston: Springer), pp. 49–61.

Some works aim at reducing dynamic and static power consumptions. The dynamic power consumption can be addressed by different kinds of techniques such as DVS (Dynamic Voltage Scaling) [6,7], AVS (Adaptive Voltage Scaling) [26], clock gating [20], asynchronous design [10] which has the particular advantage to reduce the peak power consumption. While current dynamic power optimization solutions are relevant for embedded system with high activity rates, they may be less relevant for sensor network nodes, due to their particular activity profile made of a lot of idle phases. Due to this special activity profile, solutions that address the static power consumption are also important for enhancing the system's lifetime. For that reason, static power reduction has been addressed by many works and more particularly in the sensor network context [21]. Among the techniques for reducing the static power, the use of MTCMOS (Multi-Threshold Complementary Metal Oxide Semi-conductor) technology [9] aims at reducing the leakage current with special transistors which act as power switches. Biasing solutions [22] are also proposed for saving static power.

Different works addressed the problem of power mode management. Some publications addressed the problem of the transition from active to idle state for micro-controller units [1,2]. Because this transition requires power consumption each time the system switches from an activity mode to another, it is necessary to change the power state only when necessary otherwise the gains observed by the use of a low-power mode might be counterbalanced by the losses due to transitions. Some works aim at parsing some information from the system and/or the software in order to take the best possible decisions for transiting from an active mode to a standby mode [1,2,3].

In this paper, we propose a specific hardware structure dedicated to the optimization of the transition from standby to active state. The most often used policy is to wake-up the whole system whenever a hardware event occurs. However, if this method allows the quickest possible reaction, it may be too much power consuming due to a multiplication of mode transitions, in case of unpredictable applications and/or bad transition choices from active to standby states. Actually, the "standby to active" transition requires some hardware and software applicative context knowledge. The proposed solution is a specific hardware component composed of a standard power management structure and an event driven filter in order to manage the "standby to active" transition.

The paper is organized as follows: in a first section, the architecture of interest is presented, so as the need for an adequate power control structure. A second section will be dedicated to the solution that is proposed for a new power control structure for sensor network applications. Then, the performance results are exposed and compared to a standard solution.

Related work - wireless sensor node architecture

Many sensor network platforms were proposed recently, some are commercial platforms and some are academic versions. Most of the mote platforms are based on general purpose low power micro-controllers (e.g. Texas Instruments MSP430 [8,15,13], Atmel ATMega128 [12] or Microchip PIC18xx [14]).

Then, some platforms appeared, based on a System-on-Chip (SoC) design. These architectures offer more flexibility in terms of multiple supply voltages and clock domains while simplifying the integration of analog parts. Some manufacturers already design SoC integrated platforms for sensor networks, such as Dust Networks SmartMesh-XD [16], Jennic JN513x [17], or Sensoria EnRoute500 [18].

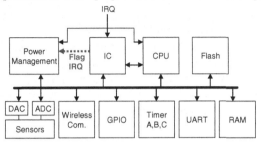

Fig. 1. Architecture of a wireless sensor node

The standard architecture considered in this paper belongs to the SoC family design and is illustrated on figure 1. In general, the digital part of the platform under study is composed of a processor core, an interrupt controller, a volatile memory block (RAM), a non volatile memory block (Flash), a General Purpose Input/Output (GPIO), several timers, a Universal Asynchronous Receiver Transmitter (UART), and an interconnect bus. The sensor part of the platform includes an Analog to Digital Converter (ADC), a Digital to Analog Converter (DAC), and the set of sensors. The communication part is constituted of a wireless communication component. Then, a power supply and clock supply block is used for powering components and driving the clocks of the platform.

The component dedicated to power management has different functionalities depending on the processor power state (i.e. "on" or "off"). During processor "on" state modes, the power manager component operates the Dynamic Power Management [24] (DPM) strategies selected by a software part by driving the supply voltage and the different clocks necessary for the system. We call this function the *"driving mechanism"*. During processor "off" state modes, the power manager component is in charge of the whole system wake-up, which is driven by hardware events (e.g. timer events or communication events) which play the role of the *"trigger mechanism"*. Most of the time, hardware events are interrupt requests (IRQ) and they are centralized by the interrupt controller which, in this case, plays the role of the trigger mechanism. Indeed, the power manager component incorporates a *"control mechanism"* that operates the reactivation of the whole system, ensuring that the wake-up procedures of components are respected.

For basic power managers, the trigger and control mechanisms are not optimal for the "standby to active" transition. In this basic approach, when a hardware event occurs, the power manager component strategy consists in activating the whole system. As it will be shown in this paper, this policy is not always optimal for decreasing the power consumption in case of low activity sensor network applications. Therefore, it is necessary to propose a solution compatible with either SoC or off-the-shelf designs of wireless platforms.

Proposed solution – Power-On Controller

The proposed solution is called Power-on-Controller (POC). The main idea of this work consists in using additional relevant information to the power management component. This information will be used for improving the flexibility of the reactivation mechanism. This section aims at describing the Power-On Controller (POC) constructed for adequately using this additional information. For describing the POC, first a description of the conception philosophy will be done. Then, a second part will describe how the POC structure has been implemented.

Inadequacy of the IC related information for power re-activation

The POC has been designed from the following observation. In previous state-of-the-art structures, IRQs are used for actuating the interrupt mechanism, which is the primary function, but they are also used as wake-up events in case the platform is in a low-power state, which is an extended function. To our opinion, the Interrupt Controller is no valuable structure for controlling the platform power management. Indeed, in platforms that use power control, the IRQs are above all dedicated and designed for interrupt processing purposes. So, a clearer identification of functions of interrupt processing and of power control is necessary in order to propose optimal hardware structure for power-on control, rather than using the structure initially dedicated to interrupt management.

So, on the one hand, the interrupt control aims at controlling the interrupt mechanism of the processor, which consists in stopping the normal execution of the processor in order to run a specific routine called the handler. On the other hand, the power-on structure aims at operating a total or partial platform reactivation consecutively to events observed on the platform. There are similarities between the two mechanisms, since they are both consecutive to the occurrence of an hardware event. For that reason, they are often based on the same information, which is presented in the last column of the table 1.

Table 1 presents the relevant information for applying adequate reaction for the interrupt control mechanism on the one hand and for the power-on mechanism on the other hand. It appears that the interrupt control related information is only a second best for being used as power-on information. In general, the interrupt masking is used as-is and the priority information is ignored and some very simple hardware is used instead, which is based on OR gates. This simple hardware allows a basic strategy of reactivating the whole platform immediately when any unmasked IRQ is detected.

The proposed POC component uses the most relevant information for processing the reactivation, which corresponds to the first column of table 1. As a consequence, the POC takes IRQ as inputs, so as the usual interrupt controller, as illustrated on figure 2. The specific information is used by the POC for applying advanced power management policies such as event accumulation and domain selection at reactivation while being easy to integrate in any SoC node platform. It should be noticed that the POC can been seen as an IC extension as well, but for simplifying the presentation, it appeared to be clearer to give a specific name to this extension. Another reason why the POC can be understood like a component clearly separated from the IC is that it is generic and it can be used within any platform. However, this adaptation would require some effort.

Table 1. Comparison of the relevant information for the interrupt mechanism and the power-on mechanism

	Power-on control	Interrupt control
Events	IRQ	
Masking Policy	Enabling/disabling the reactivation transition	Enabling/disabling the CPU interrupts
Priority Policy	Information of the urgency of the reaction (defines if an additional latency can be tolerated or not)	Information of the priority level of the HW event (helps defining which IRQ will get the CPU resource first)
Necessary HW	Information about the platform parts to be reactivated for a given IRQ	Unspecified

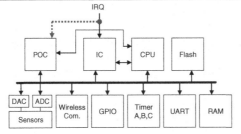

Fig. 2. POC integration in a standard wireless node architecture

Event accumulation principle:

Every time a reactivation occurs, it is associated to an energetic cost. For that reason, switching too often from a power mode to another is highly energy consuming. The additional knowledge of the urgency level of events, which is proposed with the POC, allows to apply a delayed wake-up in case the routine associated with the IRQ can tolerate any latency. This leads to event accumulation which is the fact of buffering the events which can be processed with additional delay without altering the quality of service. When an event is considered as non urgent, it is stored until an urgent one occurs, or until an internal time-out expires. Then, the system is powered-on and all pending IRQs are processed by the CPU. The energy related to the standby to active transition is consumed only once. When an urgent event occurs, it immediately triggers the reactivation transition.

Of course, event accumulation adds delay for processing non urgent events, but when events are labelled as urgent, there is no additional latency. In that sense, the quality of service remains optimal for urgent events. The event urgency labelling is done by low-level software.

Domain selection at reactivation

Another additional knowledge handled by the POC is the information about the platform parts to be reactivated for a given IRQ, as indicated in table 1. This information is used by the control part for triggering an adequate reaction when some hardware events occur. The goal is to select in advance the power supply and clock domains that will have to be reactivated when an actual reactivation event occurs.

This allows a finer control than a total reactivation, which is the only possibility on any solutions based on interrupt control information only. By doing this way, the parts that are reactivated for the event processing are in accordance with a specific event. More details about this mechanism will be given in the results section.

Flexible integration

Since the POC component does not rely on interrupt related information such as interrupt masking or priority, the component can be integrated in any standard architecture with a processor core associated to an interrupt controller (as is the case for almost all sensor network platforms).

However, it should be noticed that the event buffering capability of the interrupt controller might need to be enhanced. The enhancement could focus on the increase of event buffers sizes or on the implementation of a "buffer full" flag from the interrupt controller to the POC which would give the order to reactivate the platform.

Implementation of the POC at system level

In this section, the integration of the POC at system level is described in order to enhance the power control for peripherals and for the processor core.

Power control of peripherals

The power management of peripherals is done as follows and is summed up in table 2. During processor "off" states, the POC takes full control on the peripherals' power modes. During processor "on" states, the power management of peripherals is managed by low level software (which can be called DPM, for Dynamic Power Management) running on the processor and the POC only applies the orders of the DPM software part.

Table 2. Mechanisms involved for the power control of peripherals and processor core

	Trigger	Control	Driving
Processor "On" state	SW	SW	POC
Processor "Off" state	POC	POC	

Power control of the processor

For the processor case, the driving mechanism is operated by the POC. The control mechanism for power-on feature is done by a special FSM (Finite State Machine) that runs in the POC. This control task cannot be ensured by the processor itself because the processor may be unavailable (when "off"). Then, the trigger mechanism is ensured either by the interrupt controller or the POC. This organization is summed up on table 2.

Implementation of the POC at component level

The system view of the POC has been presented in the previous section. The current section presents the hardware structures that will apply the new policies of event accumulation and spatial selectivity for wake-up (which have been previously presented).

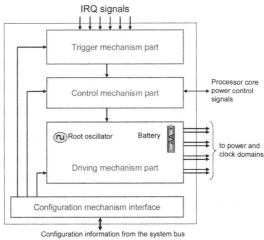

Fig. 3. Hardware structure representation of the POC

The schematic of the POC is presented on figure 3. The subcomponents are as follows.

- The first subcomponent is the driving mechanism part, which consists in the PWM (Pulse Width Modulation) signal for the supply voltage generation by the adequate analog structures such as "Buck" DC/DC converters (Direct Current/Direct Current) [11]. The generation of the different clock outputs is realized with a root PLL-based oscillator structure. Then clocks and supply voltages are routed to the whole platform.

- The second subcomponent of the POC is the trigger mechanism part. First, this part determines whether a platform reactivation is necessary or not, with or without any tolerance on the reactivation delay. Second, the trigger part defines the partial system reactivation policies to use according to the received events. The inputs of this block are the IRQ lines driven by the peripherals. In our implementation, these IRQ lines are the same as those used for standard interrupt control.

- Third, the *configuration mechanism interface* is connected to the system bus for the configuration phases. This configuration interface is located between the component and the processor. In our implementation, the configuration interface is Wishbone [4] compliant. This interface is accessed with dedicated low-level software for adequately configuring the system.
- The last part of the POC is the *control part* which manages the power mode of the processor core and peripherals. The module controls the driving part of the peripherals and of the processor core according to the external events, to the processor power control signals and to the configuration interface. It has been implemented using a Finite State Machine (FSM).

Experimental results

For evaluating the structures proposed by the POC, a realistic applicative context has been considered. According to this scenario, two approaches are compared. The first approach is a prior state-of-the-art solution such as what can be found in most of commercial or academic sensor network platforms. The results exposed here are compared to a platform that would use the Texas Instrument MSP430 microcontroller wake-up structure. It operates an immediate and total reactivation of the chip each time an unmasked event is detected. The second approach uses the POC enhancements for lowering power consumption.

The chosen scenario takes place in a sanatorium. Some patients are equipped with a bracelet which monitors their heart rate activity (this application is called "Heart monitoring") and sends alerts when a problem occurs and allows them to call for some help when needed ("Help button"). Also, the bracelets are supposed to send global running information about embedded sensors at regular intervals of 10 minutes ("Execution report"). Beside all this, a battery test is done hourly for ensuring that the bracelet never runs out of power at a bad time ("Battery test"). Table 3 shows four applications running on a given sensor node of the network and specifies the frequency of hardware events and the urgency level of this event. Table 3 also shows which hardware component is in charge of generating the event.

Table 3. Mechanisms involved for the power control of peripherals and processor for the sanatorium scenario

Application	Triggering Hardware	Event frequency	Event urgency level
Heart monitoring	Sensor part	Punctual (rare)	Urgent
Execution report	Timer	1 event / 10 min	Non urgent
Help button	GPIO	Punctual (once a day)	Urgent
Battery test	Timer	1 event / 1 hour	Non urgent

Results for event accumulation

Figure 4a shows the action of event accumulation. One can see that the number of events processed at the occurrence of a reactivation phase is superior or equal to one, while this number is one for a classic method. The curve is floored at 1 for lower values of the accumulation period that defines the maximum waiting period in case that no urgent event occurs for emptying the event queue. When the accumulation timer period is set above a particular value which depends on the application, the average number of events accumulated becomes strictly superior to one. For our application this number grows linearly for time interval configurations superior to 5 minutes.

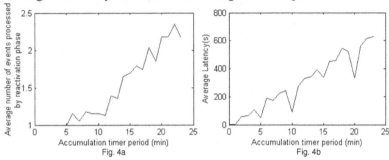

Fig. 4a   Fig. 4b

Fig. 4a and 4b. Average number of events processed for one reactivation phase and average latency

The average delay between the reception time of an event and its processing time is shown on figure 4b. Logically, the average delay for processing events grows approximately linearly with the accumulation timer period. The irregularities that appear on the curve are due to the applicative scenario where events are generated periodically. When the applicative process meets particularly well the accumulation timer it has for consequence to reduce the average delays. From this observation, we see that a good knowledge of the application can lead to keeping the average latency at reasonable levels while significantly reducing power consumption, and without impacting the delay at all for urgent signals since the average delay for processing urgent events remains approximately the same for both solutions.

Figure 5 and table 4 present the average power consumption due to events processing for the basic solution on the one hand and with the use of the POC component on the other hand. Here, significant dynamic power savings can be highlighted up to 58% for 20 minutes of timer period accumulation. To our opinion, these results show that the gains compensate the POC's consumption overhead.

Table 4. Dynamic power reduction results

Accumulation timer period (min)	5	10	15	20
Dynamic power reduction (%)	1.96	17.65	44.12	58.82

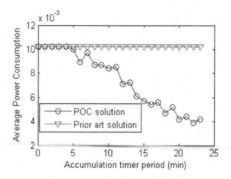

Fig. 5. Average power consumption due events handling

Results for spatial selectivity

In this second typical use case, the platform considered has distinct physical domains (i.e. clock and power supply domains) that can be handled separately. The different components of the platforms are assigned to physical domains as follows.

- **Domain – Power and power control.** It includes the POC component and the batteries. This domain is always active and allows the operation of the other domains.
- **Domain – Monitoring.** The components in this domain are able to initiate events that lead to reactivation. In order to enhance the control granularity this domains is split into 3 sub-domains:
- Sub-domain – Timer Monitoring
- Sub-domain – ADC/Sensor Monitoring
- Sub-domain – Wireless Communication Monitoring
- **Domain – Processing.** This domain includes the minimum blocks needed for processing ; that is the processor core, the interrupt controller, the volatile memory, and the interconnect bus.
- **Domain – Internal Communication.** It includes blocks such as a Universal Synchronous and Asynchronous Receiver Transceiver module (USART), a Serial Peripheral Interface controller (SPI) or an Inter Integrated Circuit bus (I^2C). This domain is activated for allowing a communication with additional peripherals.
- **Domain – External Communication.** This domain includes the modules used for communicating with other network entities. In our applicative context, it is composed of the wireless communication hardware.
- **Domain – Storage.** This domain is activated when access to storage structures is needed. Typically, non volatile memory elements such as flash memory elements are parts of this domain.

We define a mode as a combination of domain states. We consider seven modes using the domains defined above:
- **"Timer monitoring"** – One or several timers are activated for monitoring purpose
- **"Sensor Monitoring"** – ADC/Sensor is activated for monitoring purpose

- **"Wireless Monitoring"** – A part of the wireless communication is activated for monitoring purpose.
- **"Recording"** – Storing information from the processor core to the non volatile memory.
- **"Active Processing"** – The processor core is active, while no peripheral is active.
- **"Active Se"** – The sensor and processing domains are active.
- **"Active Wi"** – Wireless communication is active.
- **"Active Wi + Se"** – Wireless communication and Sensor domains are active

Table 5. Platform modes and domains

	Power and Power Control	Timer Monitoring	ADC / Sensor Monitoring	Wireless Communication Monitoring	Processing	Internal Com.	External Com.	Storage
Timer monitoring	on	on	off	off	off	off	off	off
Sensor Monitoring	on	off	on	off	off	off	off	off
Wireless Monitoring	on	off	off	on	off	off	off	off
Recording	on	off	off	off	on	off	off	on
Active Processing	on	off	off	off	on	off	off	off
Active Se	on	off	on	off	on	on	off	off
Active Wi	on	off	off	on	on	on	on	off
Active Wi+Se	on	off	on	on	on	on	on	off

Table 5 presents the organization of the platform modes according to the structural domains. This organization is compatible with SoC based approaches. In the applicative context defined above, different monitoring modes can be activated simultaneously in order to trigger events from the ADC/sensor, wireless communication, timer monitoring or GPIO. In previous solutions, when an event occurs all the systems is activated. With the POC, partial activation strategies can be defined. For example when a wireless event is raised; the POC can select the "Active Wi" mode in order to process the communication information corresponding to the event.

The overhead of the structure varies as a function of two parameters: the number of domains ($Ndom$) and the number of IRQ ($Nirq$). The impact of these parameters on the POC area has been evaluated for each POC sub component. The area of the control part is approximately constant with a variation of $Nirq$ or $Ndom$. The area of the driving part is approximately proportional to $Ndom$ and does not depend on $Nirq$. The trigger part divides into two parts. One grows proportionally to $Nirq$, and does not depend on $Ndom$. The other part of the trigger is proportional to the product of $Ndom$ and $Nirq$. Then, the configuration component is constituted of: one constant area part, one that grows proportionally to the product of $Nirq$ and $Ndom$, one that grows proportionally to $Nirq$, and a last one that grows proportionally to $Ndom$.

To sum up, the area occupation of the POC grows linearly with the two main parameters $Nirq$ and $Ndom$. This area overhead has been evaluated for an instance of the POC with $Nirq=16$ and $Ndom=4$ and the corresponding results are presented on table 6. The evaluation was performed with Synopsys Design Compiler tools [27] with a 90 nm CMOS standard cells library technology.

Table 6. Overhead due to the POC for 16 IRQ and 4 clock/voltage domains

	Trigger	Control	Driving	Configuration	Total
Number of equivalent gates	477	732	268	312	1789
Silicon Area (μm^2)	2860	4393	1607	1869	10729

Conclusion

The proposed component addresses the problem of the lifetime of autonomous nodes in WSN applicative context. WSN applications are generally less restrictive in terms of quality of service and present a low activity rate. The Power-On Controller (POC) takes advantage of these properties in order to save dynamic power consumption. The proposed solution interacts with IRQs and system core components for applying adequate event accumulation and spatial selection mechanisms.

The event accumulation mechanism proposes to limit the energy cost due to the system on chip reactivations. Indeed, system reactivation is an important power-consuming source. Event accumulation mechanism succeeds to save dynamic power consumption up to 58% on the proposed WSN application. Moreover, the POC proposes a spatial selectivity mechanism in order to wake-up hardware parts that are relevant for a given IRQ. Indeed, previous state-of-the-art solutions generally wake-up the whole system without considering the source of IRQ. The POC integration has an area overhead of 10729 μm^2 for the 90nm CMOS technology and its gate usage is 1789 gates.

It has been focused in section IV that the periodicity that may be observed on typical sensor network applicative context, such as the one described in table 3, could be exploited to properly configure the accumulation period of the POC in order to keep the average latency at reasonable levels even for non urgent events. Future work will be realized on low-level software for properly tuning the accumulation period parameter with an adaptive methodology and/or advanced applicative knowledge.

Reference

[1] Simunic, T.; Benini, L.; Glynn, P.; De Micheli, G., "Event-driven power management," *Computer-Aided Design of Integrated Circuits and Systems, IEEE Transactions on*, vol.20, no.7, pp.840-857, Jul 2001
[2] Chi-Hong Hwang; Wu, A.C.-H., "A predictive system shutdown method for energy saving of event-driven computation," *Computer-Aided Design, 1997. Digest of Technical Papers., 1997 IEEE/ACM International Conference on*, vol., no., pp.28-32, 9-13 Nov 1997
[3] Bellosa, F., "The Benefits of Event-Driven Energy Accounting in Power-Sensitive Systems", *Proceeding of the 9th workshop on ACM SIGOPS European workshop*, pages 37-42, 2000.
[4] Wishbone System-on-Chip (SoC) Interconnection Architecture for Portable IP Cores Revision B.1, Silicore Corporation, 2001.

[5] Roundy, S.; Steingart, D.; Frechette, L.; Wright, P. and Rabaey, J., "Power Sources for Wireless Sensor Networks". *In Wireless Sensor Networks*. Springer Berlin / Deidelberg, vol. 2920, pages 1-17, 2004.

[6] Chandrakasan, A.P.; Sheng, S.; Brodersen, R.W., "Low-power CMOS digital design," *Solid-State Circuits, IEEE Journal of*, vol.27, no.4, pp.473-484, Apr 1992

[7] Chandrakasan, A.; Min, R.; Bhardwaj, M.; Cho, S.; Wang, A., "Power aware wireless microsensor systems," *Solid-State Circuits Conference, 2002. ESSCIRC 2002. Proceedings of the 28th European*, vol., no., pp. 47-54, 24-26 Sept. 2002

[8] Texas Instruments, "MSP430x1xx Family User's Guide, Rev. F", 2006.

[9] Calhoun, B.; Honore, F.; Chandrakasan, A., "Design Methodology for Fine-Grained Leakage Control in MTCMOS", *2003 International Symposium on Low Power Electronics and Design, Proceedings of the*, ACM, pages 104-109, 2003.

[10] Ammar, Y.; Buhring, A.; Marzencki, M.; Charlot, B.; Basrour, S.; Matou, K.; Renaudin, M., "Wireless Sensor Network node with asynchronous architecture and vibration harvesting micro power generator", *2005 joint conference on Smart objects and ambient intelligence: innovative context-aware services: usages and technologies , Proceedings of the*, pages 287-292 , 2005.

[11] Mohan, Undeland, Robbins, "Power Electronics: Converters, Applications and Design", Wiley, 1989.

[12] Crossbow. http://www.xbow.com/

[13] Moteiv. http://www.moteiv.com/

[14] Convergix. http://www.convergent-electronics.com

[15] Dr.-Ing. habil. Jochen H. Schiller, Dr.-Ing. Achim Liers, Dr.-Ing. Harmut Ritter. Scatterweb at Free University of Berlin. Institute of Computer Science, Freie Universität Berlin, Germany. Website: http://www.inf.fu-berlin.de/inst/ag-tech/scatterweb_net/index.html

[16] Dust Networks. http://www.dustnetworks.com

[17] Jennic. http://www.jennic.com

[18] Sensoria. http://www.sensoria.com/

[19] D. Estrin et al. "Embedded, Everywhere: A Research Agenda for Networked Systems of Embedded Computers" National Research Council Report, 2001.

[20] Raghunathan, A.; Dey, S.; Jha, N.K., "Register transfer level power optimization with emphasis on glitch analysis and reduction" *Computer-Aided Design of Integrated Circuits and Systems, IEEE Transactions on*, vol.18, no.8, pp.1114-1131, Aug 1999

[21] Hempstead, M.; Wei, G.-Y.; Brooks, D., "Architecture and Circuit Techniques for Low-Throughput, Energy-Constrained Systems across Technology Generations" *Compilers, Architecture, and Synthesis for Embedded Systems (CASES), IEEE/ACM International Conference on*, October 2006

[22] Kesharvarzi, A.; Narenda, S. et. al., "Technology scaling behavior of optimum reverse body bias for leakage power reduction in Ics," *Low Power Electronics and Design, Intl. Symp. on*, pp. 252-254, 1999.

[23] Microchip Technology Inc., "Pic18f2585/4585/2680/4680 data sheet." [Online]. Available: http://ww1.microchip.com/downloads/en/DeviceDoc/39625b.pdf

[24] Benini, L.; Bogliolo, A.; De Micheli, G., "A Survey of Design Techniques for System-Level Dynamic Power Management" *IEEE Trans. on VLSI Systems*, Feb.2000. pp.299-316.

[25] Zeitzoff, P.M.; Chung, J.E., "A perspective from the 2003 ITRS: MOSFET scaling trends, challenges, and potential solutions" *Circuits and Devices Magazine*, IEEE, vol.21, no.1, pp. 4-15, Jan.-Feb. 2005.

[26] Elgebaly, M.; Sachdev, M.,"Variation-Aware Adaptive Voltage Scaling System" *Very Large Scale Integration (VLSI) Systems, IEEE Transactions on* , vol.15, no.5, pp.560-571, May 2007.

[27] Design Compiler Reference Manual, Synopsys, V.11, 2000.

Cooperation Mechanism Taxonomy for Wireless Sensor and Actor Networks

Erica Ruiz-Ibarra, Luis Villasenor-Gonzalez

{cruiz, luisvi}@cicese.mx

CICESE Research Center, Department of Electronics and Telecommunications

Km. 107 Carretera Tijuana-Ensenada, Ensenada, México 22860

Abstract. A Wireless Sensor and Actor Network (WSAN) is composed of sensor and actor nodes distributed in a geographic area of interest; the sensors are involved in monitoring the physical environment, while the actors can execute a designated task in accordance to the data collected and reported by the sensors during an event. To achieve a balanced performance, a WSAN architecture must implement an efficient cooperative communication strategy to allow the nodes to collaborate in the optimal assignment of resources and to execute tasks with the lowest possible delay. Such collaboration must take place by exchanging information and generating negotiated decisions while trying to extend the WSAN lifetime. The main contribution of this work is the proposal of a coordination mechanism taxonomy for WSANs; this taxonomy provides a framework for the classification of coordination mechanisms designed for WSAN environments. Based on this taxonomy, a comparative analysis is presented to study some of the most representative coordination mechanisms proposed in the area of WSANs up to this date.

Keywords: Coordination mechanism, Wireless sensor and actor networks, WSN, WSAN, Taxonomy.

1 Introduction

Recent advances in microelectronics and wireless technology have enabled the development of small size devices, which are low cost, power limited and equipped with wireless communication capabilities. This has led to the development of Wireless Sensor Networks (WSN), which are composed of hundreds of sensor nodes used to monitor multiple physical variables, such as temperature, humidity, sound, pressure, movement, vibrations, etc. [1]. A WSN can be deployed to support a large number of applications which include environmental monitoring, inventory tracking, prediction of natural disasters (e.g. earthquakes, forest fires), home automation, traffic control and military supervision in the battlefield [1]. However, there are some complex scenarios that require the cooperation between sensors and higher capability devices, such as actor nodes, to support the proper execution of specific tasks; wireless sensor and actor networks (WSAN) have been proposed as an important extension of WSN [2]. A WSAN can be deployed for a great variety of applications, such as microclimate control in a building or a greenhouse, detection of biological, chemical or nuclear attacks, automation of industrial processes, control of ventilation systems and heating [3].

Please use the following format when citing this chapter:

Ruiz-Ibarra, E. and Villasenor-Gonzalez, L., 2008, in IFIP International Federation for Information Processing, Volume 264; Wireless Sensor and Actor Networks II; Ali Miri; (Boston: Springer), pp. 62–73.

A WSAN is composed of a large amount of sensor nodes and a few actors connected by wireless means; these devices cooperate among themselves to provide distributed sensing and to execute specific tasks [4]. Sensor and actor nodes can be spread in the field, while a sink node can be used to monitor the network and may be used to communicate with a task manager, as illustrated in Figure 1. In a WSAN the sensor nodes behave as passive elements, collecting information from the physical world, whereas the actors are active elements that make independent decisions and are capable of executing appropriate actions in accordance to the information collected; all these capabilities allow the user to monitor and to act when located in a remote location [5]. In a WSN scenario, it is usually assumed that sensor nodes cannot be locally configured or recharged while deployed in the field, therefore this type of devices are required to be autonomous and energy efficient. The energy constraint of the sensor nodes impose limitations on the size of the device, similarly, there is a reduction of resources like memory, processing speed, computing power and bandwidth [1]; all of this with the purpose of extending the lifetime of the sensor node. On the other hand, in a WSAN, the actor nodes are equipped with greater resources, such as increased computing capacity, powerful transmitters and increased battery lifetime by means of rechargeable or replaceable power sources [5].

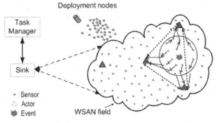

Fig. 1 Physical architecture of a WSAN [5]

The main objective of this work is to provide a common framework for the study and analysis of cooperative mechanisms in a WSAN environment. The remainder of this paper is organized as follows. Section 2 presents a description of the WSAN network architecture. Section 3 outlines a novel taxonomy for the study and the analysis of coordination mechanisms in a WSAN environment. Section 4 provides a classification and a comparative analysis of the WSAN coordination mechanisms by employing the taxonomy introduced in section 3. Finally section 5 states the conclusions of this work.

2   WSAN Architecture

A WSAN is a distributed system that can adapt and react to the environmental conditions which are reported by the collaborative effort of all the sensors and actors [6, 7]. Two different types of architectures can be defined according to the way data is collected by the sensor nodes and is reported back to the actor nodes; these are defined as the Automated and Semi-automated architectures, illustrated in Figure 2. In the Automated architecture, data is collected by the sensor nodes and it is transmitted directly to the actors, which will efficiently coordinate to execute a specific task without collaboration from the sink. As a result, automated WSAN architectures are recommended for time sensitive applications where a fast reaction by the actors is a critical requirement. In the Semi-automated architecture, the sensor data is transmitted to a central controller (e.g. the sink) which will process the collected data and will determine which actors must take action to execute a specific task; this is accomplished by transmitting a set of commands to the corresponding actors.

Fig. 2 a) Automated and b) Semi-automated Architecture [5]

The communication process, in a WSN environment, mainly takes place from the sensor nodes to the sink; on the contrary, the communication process in a WSAN environment can take place between the sensor and the actor nodes. Thus, a WSAN architecture requires the implementation of multiple coordination levels; these coordination levels are defined as: *Sensor-Sensor (SS)*, *Sensor-Actor (SA)* and *Actor-Actor (AA)*. The SS coordination is employed to gather information from the physical world in an effective and energy efficient way. The SA coordination is employed to report new events and to transmit the characteristics of the event from the sensors to the actors [7]; in addition, the SA coordination may also be used over the downlink (i.e. from the actor toward the sensor) to inform the sensors to proceed with specific sensing tasks. The AA coordination is required to execute a specific task while coordinating which actor nodes should respond within a certain area. The objective of these mechanisms is to coordinate the actions between the sensor and the actor nodes, while making an optimum use of the available resources and at the same time executing the required tasks within the time bound required by the application [8].

The following section presents a coordination mechanism taxonomy. This taxonomy is proposed to provide a framework for the analysis and evaluation of coordination mechanisms used in WSAN architectures.

3   A Taxonomy for WSAN Coordination Mechanism

To the best of our knowledge, there is no specific coordination mechanism taxonomy for WSAN networks. There is some related work, like Farinelli's [9], which presents a taxonomy for the coordination of multi-robot systems and is based in four levels: cooperation, knowledge, coordination and organization. However, this taxonomy only considers cooperative systems consisting of robots and a coordination protocol-based/protocol-free decision making process, but it does not include data and context sharing. Another related work by Salkhman et al. [10] presents a taxonomy for context aware collaborative systems based on the commonalities of different context-aware systems that emphasize collaboration, however this taxonomy is too broad, as it can be applied to a great variety of systems, from small augmented artifacts to large scale and highly distributed sensor/(actor) networks; as a result, the proposed structure does not provide the proper elements for the fine classification required in a WSAN coordination mechanism. The structure proposed by Salkhman is supported in three axis: Goal, Approaches and Means. A third work by Sameer et al. [11] develops a range of middleware services such as synchronization, localization, aggregation and tracking to facilitate the coordination through self-organizing networked sensor.

A new taxonomy for WSAN coordination mechanisms is proposed in this section; this proposal is inspired from the work presented by Salkhman [10] and Sameer [11]. The proposed taxonomy is divided in four sections as shows in Figure 3.

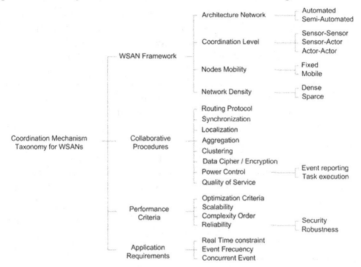

Fig. 3 Coordination Mechanism Taxonomy for WSANs

WSAN Framework. It represents the structure of a WSAN and includes:
- *Network architecture.* It is related to the way information is sensed and reported to the actors, which may be automated or semi-automated.
- *Coordination Levels.* It refers to the coordination levels employed by the coordination mechanism.
- *Node mobility.* It is used to specify if the nodes in a WSAN (i.e. sensors, actors or the sink) are mobile or fixed.
- *Network density.* It defines the ratio between the number of nodes spread throughout the field and the field dimensions.

Collaborative Procedures. It refers to those collaborative procedures between the elements of a WSAN, used to support the exchange of information by the coordination mechanism; these services include:
- *Routing Mechanism.* It is related to the procedure implemented for selecting a route to transmit packets to a destination.

- *Synchronization.* It refers to the implementation of energy-efficient techniques required to associate time and location information with the sensed data.
- *Aggregation.* It has the objective of reducing the energy consumption and the control overhead associated with the transmission of information to a common destination. To achieve this goal, a single packet is created as a result of the fusion of data which is generated by multiple sources.
- *Localization.* It is implemented to provide information regarding the geographical location of the sensor and actor nodes.
- *Clustering.* It provides a hierarchy among nodes; the substructures that are collapsed in higher levels are called cluster and there is at least one node in each cluster which is denoted as the cluster-head.
- *Power Control.* It is related to the possibility of varying the transmission power; as a result, the radio signal coverage of a node can be modified.
- *Quality of Service (QoS).* It is related to the mechanisms employed to provide guaranteed services in a WSAN architecture. In a WSAN scenario, it is possible to define two different functionalities to provide QoS support. One functionality is based on providing the required resource reservations to the nodes involved in reporting an event at the SA coordination level; in this way, it will be possible to provide differentiated services to the data being transmitted in the WSAN. A different functionality is to prioritize the execution of tasks by the actors in response to events in the WSAN. Thus, the actors should be able to respond in accordance to the priority of the events reported by the sensor nodes.
- *Data Cipher and/or Encryption.* It is related to the implementation of any cipher/encryption mechanism used to protect the integrity of data in the WSAN.

Performance Criteria. It is related to those criteria elements used to estimate the performance of the coordination mechanism; such as:

- *Optimization parameters.* This criteria denotes those metrics in which the mechanism is based to reach the proposed objectives.
- *Complexity Order.* It provides a measurement of the computational complexity of the proposed algorithms.
- *Reliability.* It provides a measure of the level of security and robustness of a WSAN coordination mechanism. To classify a coordination mechanism as secure, it must implement additional functionalities to guarantee data integrity and to avoid access to the WSAN by intruders. On the other hand, robustness is related to the capability of the coordination mechanism to handle faults, while assuring data delivery from the sensors to the actors by means of acknowledgement or retransmission procedures.

Application Requirements. The coordination mechanism must take into consideration the application requirements, such as:

- *Real-Time constraints.* It is related to the amount of time required by the coordination mechanism to report events and execute a task.
- *Event Frequency.* It is related to the periodicity of events at which the system is capable of providing a proper response.
- *Concurrent Event Support.* It is related to the capability of the coordination mechanism to support multiple simultaneous events in a WSAN.

4   Comparative Analysis

This section presents a comparison of the most relevant coordination mechanisms published up to this date in the literature. This comparison is based on the coordination mechanism taxonomy introduced in section 3; this taxonomy is divided in four sections and Tables 1- 4 show a summary of the comparative analysis for each of these. The coordination mechanism proposals are referenced using the following notation: the work presented by Melodia et al. in [12, 13] is referenced as *A*; the framework proposed by Ngai et al. [14] is referenced as *B*; the coordination mechanism provided by Yuan et al. [15] is referenced as *C*; the proposal presented by Shah et al. [16] is referenced as *D*; and the architecture developed by Melodia et al. [17] is referenced as *E*.

WSAN Framework. It represents the structure of a WSAN. The Table 1 shows the comparative analysis for the WSAN Framework.

Table 1 WSAN Framework comparison

WSAN Framework	Coordination Mechanism Approach				
	A [12, 13]	B [14]	C [15]	D [16]	E [17]
Networks Architecture					
Automated	Yes	Yes	Yes	Yes	Yes
Semi-Automated	No	No	No	Yes	No
Coordination Level					
Sensor-Sensor	No	Yes	Yes	Yes	No
Sensor-Actor	Yes	Yes	Yes	Yes	Yes
Actor-Actor	Yes	Yes	Yes	No	Yes
Node Mobility					
Sensor	Fixed	Fixed	Fixed	Fixed	Fixed
Actor	Fixed	Mobile	Mobile	Mobile	Mobile
Sink				Fixed	
Network Density	Dense Sparse	Dense	Dense	Dense	Dense

WSAN Architecture. From Table 1, it can be seen that all of the coordination mechanisms support the automated architecture given its low latency; due to the direct communication between sensors and actors. In proposal D, the authors present a coordination mechanism and evaluate it using both the automated and semi-automated architectures.

Supported Coordination Levels The coordination levels supported by each of the proposed mechanisms differ. In *A*, the SS coordination functions do not take place, as the sensors proceed to associate with an actor immediately after the detection of an event; in this way the sensor nodes will form a cluster with the actor as the cluster-head. In *B* the architecture implements clustering and data aggregation techniques between sensors, and the data corresponding to each cluster is reported to the closest actor; this is defined as SS and SA event notification, which evidently imply SS and SA coordination functions. In *C*, the proposed coordination mechanism supports the implementation of the SS, SA, and AA coordination levels. In *D*, the sensors are grouped into clusters and the average packet delay is estimated via the DAWC protocol, all these tasks are related with the SS coordination. In relation to proposal *E*, the coordination mechanism operates at the SS and AA coordination levels; it should be noticed that proposal *E* considers mobile actors and was developed as an extension of proposal *A*.

Node Mobility. The proposals summarized in Table 1, assume that the sensor nodes are fixed, while most of the proposals assume the actors to be mobile, with the exception of *A* which assumes the actors to be fixed. By considering mobile actors it is possible to consider a broader range of applications of a WSAN architecture; all this at the expense of increased complexity, as the actor mobility requires an efficient coordination.

Network Density. In relation to the network density all the proposals are capable of operating in a high density topology; however, some of the proposals do not explicitly specify this capability, as in *B*. Nonetheless it is assumed that *B* is scalable as it relies on cluster formation and data aggregation during the reporting of an event. Regarding proposal *D*, the authors explicitly describe that it can manage dense, as well as, sparse deployment of nodes.

Collaborative Procedures. Table 2 summarizes the collaborative procedures implemented in proposals *A* through *E*.

Table 2 Collaborative Procedures

Collaborative Procedures	Coordination Mechanism Approach				
	A [12, 13]	B [14]	C [15]	D [16]	E [17]
Routing Protocol	Geographical	Geographical	GAF (SS) Ad-hoc (AA)	C-DEAR D-DEAR	Geographical
Synchronization	Yes		Yes	Yes	Probably
Localization	Yes	Yes	Yes	Yes	Yes
Aggregation	Yes	Yes	Yes	NS	NS
Clustering	Event-driven	Event-driven	Hierarchical Geographical	Dynamic Weighted	
Power Control	Yes	No	No	No	Yes
Cipher/Encryption	No	No	No	No	No
QoS					
Event reporting	Yes	No	No	No	Yes
Task execution	No	No	Yes	No	Yes

Routing Support. Most of the proposed coordination mechanisms make use of geographical routing protocols, given their scalability and the fact that they can easily adapt to the location changes of the actor nodes [13, 18, 19]; in addition, it is possible to exploit the localization information of the nodes to route packets to the intended nodes in accordance with their localization within an event area. In *A*, *B* and *E* the SA coordination is based on a geographical routing paradigm, while the proposal presented in *C* makes use of the GAF [20] (Geographical Adaptive Fidelity) routing protocol at the first coordination level; in this way the sensor nodes can use the geographical localization information to transmit the data to the cluster-head, which will in turn forward the data to the closest actor. In addition, *C* proposes the use of conventional Ad Hoc routing protocol at the last level (i.e. AA coordination level), given the reduced number of actor nodes and their higher performance capabilities. On the other hand, D proposes the implementation of Delay and Energy Aware Routing protocols, such as the DEAR protocol, to transmit the data to the sink or to the actors. In the semi-automated scenario, D proposes the implementation of a centralized DEAR (C-DEAR) protocol, while for automated architectures it implements a distributed DEAR (D-DEAR) protocol; these two implementations of the DEAR protocol are designed to comply with the end-to-end delay requirements for real-time applications. In proposal E, the authors derive a simple yet optimal forwarding rule based on geographic position in presence of Rayleigh fading channels.

Synchronization. The synchronization technique is only implemented in proposals A, C and D; proposals B and E do not make any reference to the requirement of a time synchronization mechanism. In A, it is assumed that the network is synchronized by means of the implementation of an existing synchronization protocol. In C the actors will periodically transmit their geographical coordinates along with a timestamp; this information allows the cluster-head to synchronize with the actors. In D, a time synchronization mechanism is assumed to be implemented, as packet delay measurements are made by the receiver which implies that packets are tagged with timestamp information. With respect to proposal E, it does not explicitly make a reference to the requirement of a synchronization technique; however it is assumed that a time synchronization service is required, as the location estimation of the actors is made during specific time intervals.

Node Localization. The localization service is considered to be implemented by all the proposals compared. It is assumed that all the nodes are capable of knowing or determining their geographical localization; this can be implemented by means of a Global Positioning System (GPS) receiver, through trilateration techniques or any other similar approach. In proposal E, the authors make reference to a hybrid location management scheme to handle the mobility of actors with minimal energy expenditure, based on update messages sent by mobile actors to sensors.

Aggregation. The data fusion service is implemented in A, B and C, while D and E does not specify it. In A, data fusion is only implemented at the SA coordination level whenever a sensor receives information from at least two other sensors; the data is then relayed toward the actor node. In B the data fusion is implemented at the SS coordination level and further divided into different layers according to their importance; the aggregated data is then transmitted to the closet actor in the order of significance. In C data fusion is implemented at the three coordination levels, that is, the cluster-head will perform data fusion on the data received from the member nodes (i.e. fusion at the first level); at the SA coordination level all the cluster-heads associated to the same actor will construct a data-aggregation tree toward the actor; and finally at the AA coordination level, all the actors activated by a common event will construct a third data-aggregation tree toward the actor located at the center of the event area.

Clustering. The cluster formation approach is implemented in proposals A, B, C and D; on the contrary, proposal E does not rely on a cluster formation mechanism. In relation to proposals A and B, they both implement the event-driven clustering paradigm. In the event-driven clustering, the cluster formation process is triggered by an event and the clusters are created on-the-fly. Proposal C is based on a hierarchical geographical clustering paradigm, where the cluster formation is done by splitting the action area in smaller sections to create virtual grids; this strategy reduces the traffic load within each grid and makes an efficient use of resources. Finally, D proposes a Dynamic Weighted Clustering Algorithm (DAWC) which adapts to the dynamic topology of the networks; the procedure of cluster formation is not periodic and it is based on a weighting equation which sets weights to different parameters according to the application needs.

Power control. A power control mechanism is implemented in proposals A and E. Proposal A, employs a power control mechanism in the actors, which can select their power among L different levels. A higher power corresponds to a lower action completion time. On the other hand, proposal E employs a power control mechanism at the SA coordination level, where the sensors can increase the forwarding range in order to adjust the end-to end delay.

Data Cipher and/or Encryption. None of the coordination mechanisms compared make use of a security mechanism, such as ciphering or encryption, to protect the integrity of data. As a result, there is an open opportunity area for the development of coordination mechanisms which may incorporate ciphering and or encryption techniques.

Quality of Service (QoS). In relation to *Data Transfer Priorities* the proposal A, D and E implement a data transfer priority scheme. Proposal A introduces a novel notion of "reliability", which is defined as the minimum latency required by the application. To provide the required reliability, with minimum energy expenditure, A proposes a Distributed Event-driven Partitioning and Routing (DEPR) protocol, as explain in [12]. Proposal D relies on a clustering algorithm DAWC which estimates the delay budget for forwarding a packet from the cluster-heads; this algorithm guarantees the packet delivery delay to be within the given delay bound. Proposal E, similarly to A, uses the concept of "reliability", but in this case the application "reliability" requirement is achieved by adjusting the end-to-end delay by means of a power control mechanism when the traffic generated in the event area is low, and by means of a actor-driven congestion control scheme in case of congestion. Proposals C, and B are concerned with reducing the latency during the event reporting stage, during the transmission of event related information from the sensors toward the actors, but do not implement a priority scheme to support different application requirements. Regarding the *Task Assignment Priorities*, proposals C and E make use of a task priority assignment technique. In proposal C, according to the characteristic of the event, one or more actors can be triggered to perform one or more task. In proposal E, the task assignment process is achieved by means of a Mixed Integer Non-Linear Program [21] (MINLP), where the event is characterized by a tuple that describes the event characteristics, in addition it includes an event preemption policy for multi-actor task allocation for cases where resources are insufficient to accomplish a high priority task.

Performance Criteria. Table 3 shows a summary of the comparative analysis for the performance criteria.

Optimization Criteria. All of the proposals make use of the energy and latency as optimization metrics; in A the maximum allowed latency is defined as reliability. In addition to these parameters, proposals A and E measure the packet loss rate.

Scalability. Proposals *A, B, C* and *E* make use of a geographical routing protocol which is becoming the most promising scalable solutions for critically energy-constrained sensor networks [18, 19]. On the other hand, proposals *A, B, C* and *D* make use of clustering schemes which promote scalability and an efficient use of energy in the network.

Table 3 Performance Criteria comparison

Performance Criteria	Coordination Mechanism Approach				
	A [12, 13]	B [14]	C [15]	D [16]	E [17]
Optimization criteria	Energy Latency Loss packet	Energy Latency	Energy Latency	Energy Latency	Energy Latency Lost packet
Scalability	Yes	Yes	Yes	Yes	Yes
Complexity Order	-	-	-	-	-
Reliability					
Security	No	No	No	No	No
Robustness	No	No	No	No	No

Complexity Order. None of the proposals provide a clear analysis regarding the memory and the computational resources required to support the proposed coordination mechanism. In relation to *D* and *E*, they provide a measure of the order of complexity for some aspects of the coordination mechanism. However, none of the proposals provide a complete analysis to help determine the complete order of complexity of the coordination mechanism as a whole.

Reliability. It is related to the security and the robustness of the coordination mechanism. With respect to *security*, none of the proposals make use of a procedure to guarantee data integrity, or avoid network access to intruders. Regarding *robustness*, none of the proposals implement, or propose, a procedure to guarantee fault tolerance. Some proposals only measure the packet loss rate and show how the algorithms, implemented to reduce latency, also help to reduce the packet loss; however, no procedure is proposed to guarantee packet delivery through the use of acknowledgement messages or the retransmission of information.

Application Requirements. The coordination mechanism must take into consideration the application requirements. Table 4 shows a summary of the comparative analysis of the application requirements.

Table 4 Application Requirements comparison

Application Requirements	Coordination Mechanism Approach				
	A [12, 13]	B [14]	C [15]	D [16]	E [17]
Real time constraint	Low Latency	Low Latency	Yes	Yes	Low Latency
Event Frequency	Low	Low	High	High	High
Concurrent events	No	No	Yes	Yes	Yes

Real-Time constraints. In general, the proposed coordination mechanisms support real-time applications, as they promote latency reduction during the event reporting process and the execution of tasks. In relation to proposals *A, B* and *E*, they introduce a network configuration delay due to the event-based clustering approach they implement; nonetheless, these proposals try to reduce the event reporting latency. As a result, these proposals may not be suitable for real-time applications.

Frequency of Events and Concurrent event support. With respect to the frequency of events that can be handled by the coordination mechanism, proposals *A* and *B* are not capable of providing an efficient operation under high frequency event scenarios, this is also true for scenarios involving multiple events; this is related to the limitations imposed by the event-driven clustering paradigm. In relation to proposals *C*, *D* and *E*, they are able to operate in frequent-event scenarios, as well as, scenarios involving multiple simultaneous events.

5. Conclusions

This work analyzes the different cooperation strategies implemented at the nodes of a WSAN, which are used to achieve distributed sensing and the execution of tasks in accordance to the sensed data. One of the main contributions of this work is the proposal of coordination mechanism taxonomy for WSAN architectures; the taxonomy provides a framework suited to the specific requirements of a coordination mechanism designed for a WSAN environment. The proposed taxonomy is divided in four sections: WSAN Framework, Collaborative Procedures, Performance Criteria and Application Requirements.

In addition to the taxonomy, a comparative analysis of the most representative coordination mechanism, published in the literature up to this date, has been provided. In general, the proposed mechanisms proceed to split the event area and perform a hierarchical coordination by employing localization information; this is done with the objective of selecting the proper nodes (i.e. sensors and actors) that will react in response to a specific event with the smallest possible response time. The proposed applications for each of the coordination mechanisms differ with respect to the frequency of the events. In general, the proposed coordination mechanisms try to comply with the support of real-time response requirement along with an efficient use of energy in the WSAN. On a final note, none of the coordination mechanisms analyzed implement a mechanism that guarantee the security of the data and system robustness.

References

1. Romer, Kay, Mattern F: The Design Space of Wireless Sensor Networks. IEEE Wireless Communications, (2004).
2. Vassis D, Kormentzas G, Skianis C: Performance evaluation of single and multi-channel actuator to actuator communication for wireless sensor actuator networks. Ad Hoc Networks, Elsevier Science, (2006).
3. Petriu EM, Georganas ND, Petriu DC, Makrakis D, Groza VZ: Sensor-based information appliances. IEEE Instrumentation and Measurement Magazine (2000).
4. Akyildiz Ian F, Su W, Sankarasubramaniam Y, Cayirci E: A Survey on Sensor Networks. IEEE Communications Magazine (2002).

5. Akyldiz F, Kasimoglu I: Wireless sensor and actuator networks: Research challenges. Ad Hoc Networks Journal Elsevier. (2004).
6. Cayirci E, Coplo T, Emiroglu O: Power Aware Many to Many Routing in Wireless Sensor and Actuator Networks. Proc. 2nd European Workshop on Wireless Sensor Networks, (2005).
7. Hu F, Cao X, Kumar S, Sankar K: Trustworthiness in Wireless Sensor and Actuator Networks: Toward Low-complexity Realibility and Security. IEEE Global Telecommunications Conference, GLOBECOM, (2005).
8. Díaz M, Garrido D, Llopi L, Rubio B, Troya JM.: A component framework for wireless sensor and actuator network. 11th IEEE International Conference on Emerging Technologies and Factory Automation. (2006).
9. Farinelli A, Iocchi L, Nardi D: Multirobot systems: a classification focused on coordination. Systems, Man and Cybernetics, Part B, IEEE Transactions Volume 34. (2004).
10. Salkham A, Cunningham R, Senart A, Cahill V: A Taxonomy of Collaborative Context-Aware Systems. Workshop on Ubiquitous Mobile Information and Collaboration Systems UMICS'06. (2006).
11. Sundresh S, Agha G, Mechitov K, Kim WY, Kwon YM: Coordination Services for Wireless Sensor Networks. International Workshop on Advanced Sensors, Structural Health Monitoring and Smart Structures. (2003).
12. T Melodia, D Pompili, VC Gungor, and IF Akyildiz: A distributed coordination framework for wireless sensor and actuator networks. Proc. 6th ACM international symposium on mobile ad hoc networking and computing. 2005.
13. Melodia T, Pompili D, Gungor VC, Akyildiz IF: Communication and Coordination in Wireless Sensor and Actor Networks, IEEE Transactions on Mobile Computing, (2007).
14. Ngai ECH, Lyu MR, Liu J: A Real Time Communication Framework for Wireless Sensor-Actor Networks. IEEE Aerospace Conference, (2006).
15. Yuan H, Huadong M, Hongyu L: Coordination Mechanism in Wireless Sensor and Actuator Networks. First International Multi-Symposiums on Computer and Computational Sciences - IMSCCS'06. (2006).
16. Shah, G.A, Bozyigit M, Akan OB, Baykal B: Real Time Coordination and Routing in Wireless Sensor and Actuator Networks. Proc. 6th International Conference on Next Generation Teletraffic and Wired/Wireless Advanced Networking NEW2AN (2006).
17. Melodia T, Pompili D, Akyldiz I: A Communication Architecture for Mobile Wireless Sensor and Actor Networks. In Proceeding of IEEE International Conference on Sensor, Mesh and Ad hoc Communications and Networks SECON. (2006).
18. J Li, J Jannotti, DD Couto, D Karger, and R Morris: A scalable location service for geographic ad hoc routing, in Proceedings of ACM/IEEE MobiCom 2000, Boston, Massachussets. (2000)
19. R Jain, A Puri, and R Sengupta: Geographical routing using partial information for wireless ad hoc networks, IEEE Personal Communications. (2001)
20. Y Xu, J Heidemann, and D Estrin: Geographyinformed Energy Conservation for Ad Hoc Routing, Proceedings of the 7th annual international conference on Mobile computing and networking, (2001).
21. J Czyzyk, M Mesnier, and J More: The NEOS server, IEEE Journal on Computational Science and Engineering, vol. 5. (1998).

Extending Network Life by Using Mobile Actors in Cluster-based Wireless Sensor and Actor Networks

Nauman Aslam, William Phillips, William Robertson and S. Sivakumar

Department of Engineering Mathematics & Internetworking
Dalhousie University, Halifax, Nova Scotia, Canada, B3J2X4
{naslam, william.phillips, bill.robertson,sivas}@dal.ca

Abstract

Wireless sensor actor networks (WSANs) consist of a large number of resource-constrained nodes (sensors) and a small number of powerful resource rich nodes (actors). This paper investigates the case where sensors are organized into clusters and mobile actors are used for maintaining an energy efficient topology by periodically manipulating their geographical position. We present an elegant technique that allows actor nodes to find an optimal geographical location with respect to their associated cluster heads such that the overall energy consumption is minimized. The simulation results demonstrate that the technique proposed in this paper significantly minimizes energy consumption and extends the network lifetime compared with traditional cluster-based WSN deployments.

Keywords: wireless sensor actor networks, mobile actors, energy conservation, clustering, network lifetime.

1. Introduction

Wireless Sensor Actor Networks (WSANs) have emerged as the-state-of-the-art technology in data gathering from remote locations by interacting with physical phenomena and relying on collaborative efforts by few resource rich and a large number of low cost devices [1]. In WSANs sensors are low cost resource constrained devices that have limited energy, communication and computation capabilities. Once deployed, sensor nodes collect the information of interest from their on board sensors, perform local processing of these data including quantization and compression, and forward the data to a base station (BS) through a cluster head or gateway node. Actors, on the other hand are resource rich nodes that are equipped with better possessing power, higher transmission powers, more energy resources and may contain additional capabilities such as mobility.

Despite the tremendous technological advancements in the field of wireless sensor network (WSNs), energy conservation is still one of the fundamental challenges in WSN system design. In a sensor node, energy is consumed in sensing, computation and processing, however the wireless transceiver consumes a significant amount of energy as compared to all other sources. Most of the WSN deployments are based on

Please use the following format when citing this chapter:

Aslam, N., et al., 2008, in IFIP International Federation for Information Processing, Volume 264; Wireless Sensor and Actor Networks II; Ali Miri; (Boston: Springer), pp. 74–84.

an application specific data gathering i.e. a large number of sensor nodes send their data to the BS. Clustering is employed as a standard approach for achieving energy efficiency and scalable performance in WSNs [2]. Sensor nodes can be organized hierarchically by grouping them into clusters, where the data is collected and processed locally at the cluster head. Single hop clustering protocols follow two-tier architecture, consisting of sensor nodes and cluster heads. In tier one communication, each sensor node sends it data to its local cluster head. In tier two the cluster heads send the processed data to the BS. In protocols where the cluster head selection is performed randomly, the distribution of cluster heads is not well controlled. Most protocols based on the two-tier architecture use direct communication for sending cluster head data to the BS. If a cluster head is located far away from the sink, its energy consumption would be greater than the one located closer to the sink for similar traffic patterns from cluster members. Intuitively, deployment of resource rich actor nodes could off set some of the higher transmission costs incurred by cluster heads in data transmission to the BS. In such a case, each elected cluster head will associate with an actor node which will forward cluster head's data to the BS. Since random clustering involves election of new cluster heads in each round therefore deployment of static actor nodes will not be energy efficient. Such inefficiency may result from variations in distances between the actor node and its associated cluster heads and their respective residual energy levels. The concept of mobility solves this dilemma by effective location management of actor nodes thus providing the ability for enhanced energy savings. Figure 1 illustrates a scenario where mobile actors are used for relaying cluster head's data to the BS. This paper proposes a methodology where mobile actor nodes relocate to a new location in each round such that the energy consumed by associated cluster heads is minimized. A weighted cost based function that accounts for the residual energy levels of associated cluster heads is used to find the new of location the actor nodes.

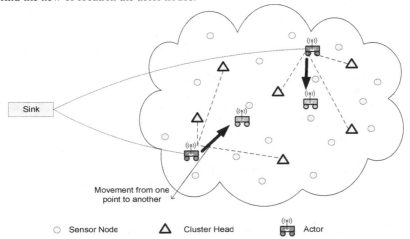

Figure 1: Example of Sensor Field with Cluster Heads and Mobile Actors

The concept of mobile nodes in WSN is relatively new and has been proposed in various contexts. One of the earliest proposed researches was concept of data mules [3]. Data mules are referred as mobile entities that collect data from sensors as they pass by the sensor nodes buffer the collected data and drop the data off to the wired access point. Data mules were primarily aimed at collection of non real-time data. Authors in [4] used mobile nodes for data collection in sparsely connected network. First nodes are partitioned into several groups with respect to their data generation rates and location. Then the schedule for each group is created using the traveling salesman problem such that there is no data loss or overflow. Message Ferrying is introduced in [5, 6] where a set mobile nodes (message ferries) provide communication service for nodes in sparse connected networks. In the message ferry approach, each node communicates only with the message ferries, therefore all communication cost is reduced to a short distance single hop communication at the expense of increased message delay. This paper adopts a rather simple approach by exploiting the actor mobility to achieve balanced energy dissipation by cluster heads. The rest of the paper is organized as follows. Section 2 presents system model and assumptions. Section 3 presents details of our proposed technique for location management of mobile actors. Simulation results are discussed in Section 4. Finally, our main conclusions and future research directions are highlighted in Section 5.

2. System Model and Assumptions

We make the following assumption for our sensor network:
1. The network consists of a large number of resource constrained nodes (referred as sensing nodes) and small number of resource rich nodes (referred as actor nodes) with enhanced communication capacity, ability to recharge their energy source and mobility.
2. We assume that all nodes have means to find out their geographical position.
3. The nodes are dispersed in a 2-dimensional space following a uniform distribution.
4. The base station is located outside the deployment region and has no energy constraints.
5. Once deployed the sensor nodes remain stationary and continue to operate until they completely exhaust their batteries.
6. We use the energy model presented in [7]. In this model the energy required to run the transmitter or receiver electronic circuitry E_{elect} is 50 pJ/bit and ε_{amp} the energy to run transmitter amplifier is 100 $pJ/bit/m^2$. The energy expended by a node in transmitting an l-bit message a distance of d is given by,

$$E_{Tx} = lE_{elect} + l\varepsilon_{amp} d^2$$

(1)

The energy expended in receiving an l-bit message is given by,

$$E_{Rx} = lE_{elect}$$

(2)

3. Location Management of Mobile Actors

In this section we present our rationale and algorithm for location management of mobile actors. The presented technique is based on the motivation that all cluster heads should have a balanced energy dissipation rate. This objective is achieved by introducing a cost function that allows the actor nodes to adjust their geographical location in such a manner that the overall energy of associated cluster heads is minimized. To further elaborate, let's consider an example illustrated in Figure 2. Four cluster head A, B, C and D are associated with an actor node. The location of each node is shown below in 'x' and 'y' coordinates. Figure 2a shows the initial location of actor node and its associated cluster heads. Since distance from each cluster clusters head to the actor node is different, it is impossible have comparable energy consumption for all cluster heads. Intuitively, finding a centroid as illustrated in Figure 2b with respect all cluster head locations would provide a position where actor node is at the minimum distance from each cluster head. However, finding a centroid would still leads to uneven energy consumption rate if the residual energy levels of all associated cluster heads are different. To account for the residual energy levels we introduce a weighted cost based functions that allows actors node to fine tune its location based on the residual energy level. Figure 2b also shows the actor node's updated location where it moves closer to the node A that has minimum residual energy. Now we describe calculation of cost factor. Table 1 outlines the notation used in the following sections.

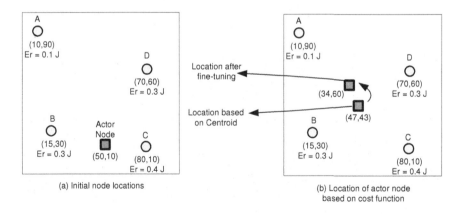

(a) Initial node locations   (b) Location of actor node based on cost function

Figure 2: Example illustrating location management of actor node using weighted cost function

Table 1: Notations Used

Set of actor nodes	A
k^{th} node in A	a_k
Target location coordinates of a_k	(X_k, Y_k)
Set of cluster heads	CH
i^{th} node in CH	ch_i
Location of ith cluster head	(x_i, y_i)
Residual energy of ith cluster head	e_i
Cost factor for i^{th} cluster head	ε_i

We assume that cluster heads are elected using any generic single hop clustering protocol such as [2, 8, 9]. Once elected, each cluster head will listen for the *Actor Advertisement Messages* (sent by actor nodes). Each cluster head then finds a least cost actor node and sends an *Association Message* to the selected actor node indicating its location coordinates and residual energy level. Upon receiving the *Association Message* the actor node knows that it will forward cluster heads data to the BS. Once the actor node receives *Association Message* from all it calculates the cost factor for each cluster head *i* as follows,

$$\varepsilon_i = \frac{(e_{max} - e_i)^2 + e^2_{min}}{(e_{max} - e_{min})^2 + e^2_i} \tag{3}$$

Where,

$$e_{max} = \max(e_i) \quad \text{for } i \in CH^k$$

$$e_{min} = \min(e_i) \quad \text{for } i \in CH^k$$

The calculated cost factors are used to find the new location coordinates X_k and Y_k from equation (4) and (5) respectively.

$$\frac{d}{dx} \sum_i \varepsilon_i (x - x_i)^2 = 0 \tag{4}$$

$$\frac{d}{dy} \sum_i \varepsilon_i (y - y_i)^2 = 0 \tag{5}$$

Based on the discussion related to Figure 2 above, each mobile actor node utilizes the cost function to optimally adjust its coordinates. This is done so by taking into account the residual energy levels of its associated cluster heads and their communication cost to the actor node. The pseudo code of the steps used for location management in cluster based environments is outlined in Algorithm-I.

Algorithm-I:

Begin

    $\forall \; ch_i \in CH$

1. Receive Actor Advertisement Messages
2. Find least cost Actor Node
3. Send Association Message

    $\forall \; a_k \in A$

4. *Send* Actor Advertisement Message
5. *While* Timer '*T*' is valid
6. Receive Association Messages from cluster heads
7. end *While*
8. Calculate weighted energy cost ε_i for all associated cluster heads
9. Calculate new X_k and Y_k
10. Move to new position (X_k , Y_k)

End

---

4. Simulation Results

This section presents the performance analysis of our proposed technique. We assume ideal conditions at the physical level such that the probability of wireless interference is negligible. The network simulation model was built using MATLAB. We used the clustering algorithm presented in [8]. We also assume that the network consists of a small number of actor nodes (ranges from 1% to 3.5% of the network size). The simulation proceeds in data collection rounds. Since each data collection round consists of cluster set up and actor node selection phase. In each round, each non cluster head node sends one data packet to its cluster head. The cluster head performs in-network processing by aggregating the data received from its member nodes and forwards this data to the actor node. Finally, the actor node forwards data received from its associated cluster head to the BS. We also assume that the actor nodes do not transmit any data while they are moving from one location to another. Mobility feature is only used to optimize the energy conservation. Two performance metrics are used. The first one is denoted by network lifetime as used in [2, 8, 9]. This metric represents network lifetime in data collection rounds from the instant the network is deployed to the moment when the first node runs out of energy. The second performance metric is the average cluster head energy consumed per round. Table 2 summarizes the important simulation parameters used.

It can be observed from Figure 3 and 4 that by using mobile actors the network lifetime is increased in more than 100% as compared to the direct transmission case. It is also worth noticing the sharp slope for mobile actor's case which is a result of balanced energy dissipation in the sensor nodes.

Table 2: Simulation Parameters

Sensor Deployment Area	100 x 100 m
BS Location	(50,175) m
Number of Nodes (N)	200 , 400
Data Packet Size	4000 $bits$
Control Packet Size	25 $bytes$
Initial Energy of Actors	20 J
Initial Energy of Sensors	0.5 J
Number of Actor Nodes	1 % to 3.5 % of N
E_{elect}	50 $n J/bit$
ε_{mp}	100 $pJ/bit/m^2$

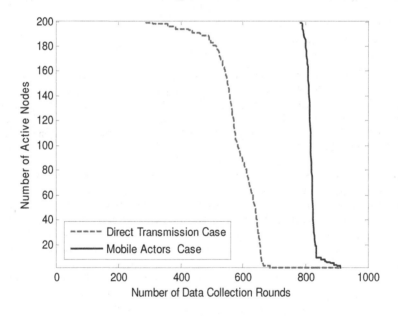

Figure 3: Network life (200 Nodes Network)

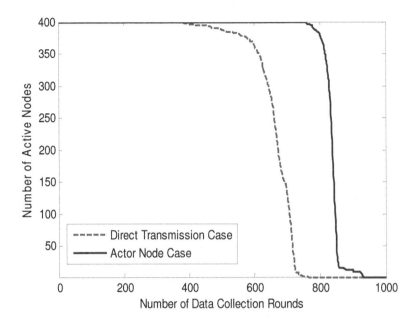

Figure 4: Network life (400 Nodes Network)

We also investigated the optimal number of actor nodes for the two network sizes of 200 and 400 nodes. Figure 5 and 6 show how network life is effected as a result of varying the number of actor nodes. The x-axis denotes the number of mobile actors in the system and the y-axis denotes the number of rounds when the first node runs out of its energy. The number of mobile actors is varied from 1% of network size to 3.5% of the network size. It can be observed from both Figure 5 and 6 that a value of 2% (4 in 200 nodes network and 8 in 400 nodes network) provides the best results. When the number of mobile actors is 1% of the network size the number of associated cluster heads per actor node is large. Hence it may lead to situations where the resultant topology cannot be further optimized. Therefore adding a small number of actor nodes further increases energy savings resulting in enhanced network lifetime. Conversely, a relatively large number of actor nodes will results in more overhead messages. Hence it is desirable to have a minimum number of actor nodes that provide optimal coverage per cluster head. From the simulations we conclude that the best value for network life time is given by when the number of mobile actors is 2% of the network size.

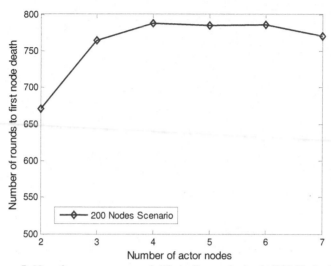

Figure 5: No. of actors vs. network life in first node death (200 Nodes Networks)

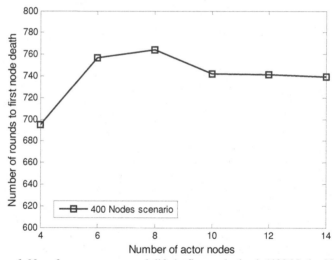

Figure 6: No. of actors vs. network life in first node death (400 Nodes Networks)

Figure 7 illustrates the mean cluster head energy consumed per round for mobile actors and direct communication scenario. The statistics are collected using 100 independent runs. It is clearly observed that the cluster heads in former case consume far less energy (almost 50%) because of shorter distance transmission to the mobile actors as compare to the direct communication case. Hence the overall effect in network life time gain is quite significant.

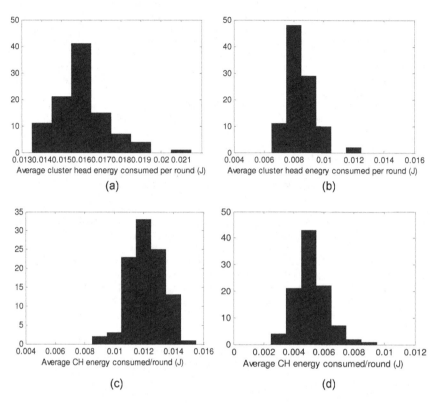

Figure 7: Histogram for average cluster head energy consumed per round (J). (a) 400 Nodes Direct communication case (b) 400 Nodes with mobile actors (c) 200 nodes with direct communication case (d) 200 Nodes with mobile actors.

5. Conclusions and Future Work

In this paper we have presented a simple and elegant technique to use mobile actor nodes for extending the network life time. We have shown that by deploying a small number of resource rich mobile nodes the network life time can enhance in excess of 100% as compared to single hop clustering. We presented a weighted cost based function that allows mobile actors to fine tune their location by considering the residual energy levels of associated cluster heads. We intend to investigate effects of different mobility models in our future research.

References

[1]   I. F. Akyildiz and I. H. Kasimoglu, "Wireless sensor and actor networks: research challenges," *Ad Hoc Networks*, vol. 2, pp. 351-367, 2004.

[2]   W. Heinzelman, A. Chandrakasan, and H. Balakrishnan, " An Application-Specific Protocol Architecture for Wireless Microsensor Networks," *IEEE Transactions on Wireless Communications*, vol. 1, pp. 660-670, 2002.

[3]   R. C. Shah, S. Roy, S. Jain, and W. Brunette, "Data MULEs: modeling and analysis of a three-tier architecture for sparse sensor networks," *Ad Hoc Networks*, vol. 1, pp. 215-233, 2003.

[4]   Y. Gu, D. Bozdag, E. Ekici, F. Ozguner, and C. G. Lee, "Partitioning based mobile element scheduling in wireless sensor networks," *Sensor and Ad Hoc Communications and Networks, 2005. IEEE SECON 2005. 2005 Second Annual IEEE Communications Society Conference on*, pp. 386-395, 2005.

[5]   W. Zhao and M. H. Ammar, "Message ferrying: proactive routing in highly-partitioned wireless ad hoc networks," *Distributed Computing Systems, 2003. FTDCS 2003. Proceedings. The Ninth IEEE Workshop on Future Trends of*, pp. 308-314, 2003.

[6]   W. Zhao, M. Ammar, and E. Zegura, "A message ferrying approach for data delivery in sparse mobile ad hoc networks," *Proceedings of the 5th ACM international symposium on Mobile ad hoc networking and computing*, pp. 187-198, 2004.

[7]   K. Dasgupta, K. Kalpakis, and P. Namjoshi, "An efficient clustering-based heuristic for data gathering and aggregation in sensor networks," *Wireless Communications and Networking, 2003. WCNC 2003. 2003 IEEE*, vol. 3, 2003.

[8]   N. Aslam, W. Robertson, S. C. Sivakumar, and W. Phillips, "Energy Efficient Cluster Formation Using Multi-Criterion Optimization for Wireless Sensor Networks," in Proc. 4th IEEE Consumer Communications and Networking Conference (CCNC), 2007.

[9]   M. Ye, C. F. Li, G. H. Chen, and J. Wu, "EECS: An Energy Efficient Clustering Scheme in Wireless Sensor Networks," *IEEE Int'l Performance Computing and Communications Conference (IPCCC)*, pp. 535-540, 2005.

Deployment-based Solution for Prolonging Network Lifetime in Sensor Networks

Sonia Hashish and Ahmed Karmouch

School of Information Technology and Engineering(SITE)
University of Ottawa-Canada
{shashish, karmouch}@site.uottawa.ca

Abstract. Enhancing sensor network lifetime is an important research topic for wireless sensor networks. In this paper, we introduce a solution for prolonging the lifetime of sensor networks that is based on a deployment strategy. In our proposal, data traffic is directed away from the network center toward the network peripheral where sinks would be initially deployed. Sinks stay stationary while collecting the data reports that travel over the network perimeter toward them. Eventually perimeter nodes would be exposed to a *peeling phenomenon* which results in partitioning one or more sinks from their one-hop neighbors. The partitioned sinks move discrete steps following the direction of the progressive peeling The mechanism maintains the network connectivity and delays the occurrence of partition. The performance of the proposed protocol is evaluated using intensive simulations. The results show the efficiency (in terms of both reliability and connectivity) of our deployment strategy with the associated data collection protocol. **Keywords**: Sensor networks, Data collection, Mobile sinks, Deployment.

1 Introduction

Maximizing the lifetime of sensor network would convert the theoretical achievements in sensor network research to successful real world deployed networks.

One of the reasons that threaten the lifetime of sensor networks is the energy depletion of nodes located around the data collector nodes (usually called sinks). Such energy depletion forms what is called an energy hole around the data collector [1]. Energy hole is most likely unavoidable in networks depend on stationary-based sinks. This is definitely true for networks with uniform node distribution and uniform data reporting. Since data travels to the collector using multi-hop communication model, nodes located near the collectors are exposed to heavier load. Such nodes lose their energy resources faster than others and an early sink-network partition occurs.

Maximizing sensor network lifetime is considered as an optimization problem in linear programming. It has been proved to be NP-hard [2]. Approximation algorithms are developed to provide efficient solutions under certain settings. The dynamicity and the frequent updates in the network topology are usually unpredictable which reduces the efficiency of such solutions in many cases. Many other practical solutions have been considered within the literature to minimize the tendency of the nodes located near sinks to form an early partition. Examples of these solutions are (1) replacing the stationary sinks with mobile ones, (2) considering the dynamic clustering of the network, (3) intentionally performing non-uniform (controlled) node

Please use the following format when citing this chapter:

Hashish, S. and Karmouch, A., 2008, in IFIP International Federation for Information Processing, Volume 264; Wireless Sensor and Actor Networks II; Ali Miri; (Boston: Springer), pp. 85–96.

distribution over the coverage area. Although effective solutions, they have some limitations and constraints in real world deployments.

Using different clustering techniques to enhance the network performance has been known for long time in the area of ad- hoc networks as well as sensor networks. However, clustering algorithms are expensive in terms of their communication cost. They require extensive overhead that could contribute to drain the resources of sensor networks [3, 4].

Exploiting sink mobility to enhance sensor network lifetime raises many new problems related to the mobility. Examples are: what is the best mobility pattern that would be implied and how to lessen the interference of the mobility trajectories (in case of multiple mobile sinks) [5]. In addition to the latency problem that results from the difference between the speed of moving the sink to collect the data versus the multi-hop transmissions of the data. Some techniques imply infrequent mobility [6]. Mobile sinks infrequently move to balance the load among nodes and a multi-hop wireless transmission is used as the main regime of data transportation.

On the other hand, intentionally performing non-uniform node distribution [7] is problematic and difficult to be achieved in real world deployments. This is due to the expected large number of sensor nodes that usually should be randomly deployed within the coverage area [8].

In this paper we introduce a solution that maximizes the sensor network lifetime. The proposed solution combines the benefits of using multiple-mobile sinks and the non-uniform node distribution (without performing non-uniform node distribution). Our proposed protocol reduces the problems associated with the above mechanisms. In our proposal, data traffic is directed away from the network center toward the network peripheral where sinks would be initially deployed. Sinks stay stationary while collecting the data reports that travel over the perimeter toward them. Eventually perimeter nodes at the network peripheral would be exposed to a *peeling phenomenon* (have their energy depleted exposing other inner nodes to be perimeter) which results in partitioning one or more sinks from their one-hop neighbors. The partitioned sinks move discrete steps following the direction of the progressive peeling and the connectivity is re-established. The protocol is totally dynamic. The overhead of exchanging topology updates messages associated with the mobility is not required. We show that our solution leads to a sub-optimal energy balancing through the network. The balance of both load and energy consumption leads to an expansion of the network lifetime. The performance of the protocol is shown by intensive simulations that consider realistic conditions of the underlying network settings. The results show the efficiency of the protocol in expanding the network lifetime, minimizing overhead associated with sink-mobility, and achieving high degree of reliability.

The rest of this paper is organized as following Section 2 discusses current related work. Section 3 explains our assumptions about the underlying network model. Section 4 describes the design details of the proposed boundary-peeling data collection protocol. Section 5 summarizes the performance of the approach. Discussions are given in section 6. Section 7 provides a brief conclusion.

2   Related work

Extensive work has been proposed into the literature to enhance the sensor network lifetime using different conceptual approaches. Due to the limited space we only list those we believe they are the most related to our work. Routing mechanisms that aim at balancing energy consumption and load through the network have been taken a considerable attention. Some of these protocols consider mobile sinks. Their main concern is how to route toward the mobile sink rather than how to exploit the mobility of sinks to enhance the lifetime of the network. A good survey of such mechanisms is [9]. The guidelines study of energy balancing that lead to enhancing the lifetime in sensor networks which consider static sinks is given in [10]. Authors conclude that the optimal energy balancing is impossible under some conditions. They also suggest a solution based on training the network into coronas with equal widths. The widths of such coronas should be determined based on the energy loss factor.

Many approximation algorithms have been developed for maximizing the lifetime of the network. Most of these algorithms are interested in the behavior of sensor node rather the than sink node. One of the proposals that consider the behavior of the sink node is found in [11]. Authors determine the location of the base station node (the data collector) that could result in maximizing the lifetime. The algorithm considers the routing strategy that would be used but no mobility is considered.

Exploiting mobility to extend network lifetime has been introduced recently into wireless sensor networks. More and more work has been devoted to address the effect of sink mobility on enhancing the performance of the network. In some proposals [12-14] mobile sinks would be used as transportation units where they move to collect the data from the stationary nodes. Such approaches result in significant energy saving but the latency problem is the main concern. Other approaches consider mobility of the base station as an assisted facility to improve the network performance [5, 6, 15]. In such assisted mobility protocols, the main regime of data transfer is the wireless multi-hop transmissions; mobility helps reducing hot spots and increases the load balancing through the nodes. So these approaches are always associated with data collection mechanisms that maximize the benefits of the mobile sinks.

Our work belongs to the assisted mobility protocols. The most relevant work to our work is that in [15]. The authors analytically prove that the optimal mobility trajectory in case of circular networks is where the mobile sink moves around the network peripheral. Our work in this paper complements such analytical foundations and inspired by our previous work in the area of data dissemination [16]. Authors in [15] suggest that the mobile sink constructs a global routing tree based on the shortest path while moving. The peripheral nodes forward their data using trajectory forwarding routing follows the mobility trajectory (in that case a circle). The scheme tends to be less adaptable to the topology irregularities. Both shortest path tree and trajectory forwarding are proactive mechanisms which require extra communication overhead to keep their performance high.

Joint mobility strategy is also suggested in [6] where a routing scheme is proposed to route the data to a single mobile sink. The sink moves around the network peripheral. Authors propose that each data collection round is preceded by a sampling round. In a sampling round, the sink stops at certain anchor positions and determines the optimal visiting time by sampling the global power consumption. The visiting time is then applied to control the movement of the sink while in the data collection round. Both [15] and [6] consider a single mobile sink. Multiple-mobile sinks raise the fact that coordination is required to keep the "positive effect" of the mobility trajectories interference on the network performance. Authors in [5] provide multiple algorithms to solve such problem.

Our work differentiates itself by considering multi-mobile sinks with mobility pattern that inherently simple and non-interfereable. The protocol efficiently adapts to the topological changes. It doesn't require any global knowledge that should be collected in a centralized manner. It also works for any network topology whether regular or irregular.

3 Network Model

Network infrastructure: We assume a two-dimensional terrain area. The distribution of the nodes within the terrain area is uniformly random. Sensor nodes have the same capabilities in communication, computation and storage. Each sensor node knows the coordinates of its location using either GPS or any existing localized techniques [17]. Two nodes are considered to be neighbors if each of them is located in the transmission range of the other. Sinks or data collectors are special nodes with no restriction in their energy resources or their communications capabilities. We assume multiple of such special nodes are initially placed arbitrary at the network peripheral. Data collector nodes could be the end-points of the data collection process or they could further forward the collected data to other remote site over long wireless radio.

Energy and traffic models: We assume that all sensors have the same initial energy rate e. Although sensors use their energy while they are transmitting, receiving and sensing, we assume that the transmission process is the dominant factor of the energy depletion [17]. We use the uniform data reporting model where each sensor i generates data with fixed rate μ_i. Each node consumes a unit of energy to transmit one unit of data. So the number of transmissions performed by each node is a good indicator of its energy consumption rate. No data aggregation is considered. Each node transmits its own data and any data units that it receives to forward on behalf of other nodes. Sensor nodes complete their task when they forward their data to any of the data collector nodes.

Mobility Model: we assume that each sink has a controlled mobility facility [18]. Sinks travel using a discrete mobility pattern M where the movement is performed in steps and only under some conditions (as we will explain in next section).

3.1 Balancing the energy expenditure

In this section we show that our protocol that we describe in next subsections leads to a sub-optimal energy balancing through the network as defined in [7]. Authors in [7] consider a circular network model where the network area A is partitioned into coronas with equal widths (fig. 1). They prove that, if the sink is located at the center of the network c and the nodes are distributed non-uniformly according to the following rule: $N_i/N_{i+1} = N_{i+1}/N_{i+2} = \ldots = q$ (where $q > 1$ and N_i is the number of nodes in area A_i known as corona i), a suboptimal energy balancing could be achieved. In other words if the number of nodes in corona i which is nearer to the sink is larger than the adjacent corona $i+1$ by a factor q a suboptimal energy balancing is achieved. The authors define the suboptimal energy balancing as the ability to balance the energy among all the inner parts of the network except the outmost one.

Distributing a large number of nodes within the coverage area using the above rule could be difficult in real world deployment. We prove that without the need for such controlled distribution, we could achieve a similar sub-optimality. This could be achieved by considering deploying the data collector nodes at the network peripheral. In addition to such exterior deployment, a data dissemination strategy that allows the direction of traffic to be reversed (from inner parts with smaller number of nodes to the outer parts with larger number of nodes) should be implied.

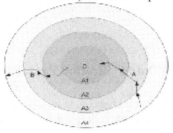

Fig. 1. Area –Dissemination direction relationship. The network is partitioned into number of coronas. Paths *A, B* represent two directions of traffic. *C* represents the network center.

In fig. 1 path *A* represents one possible data traffic paths from outer corona to inner one where the sink is assumed to be located at the network center. Non-uniform node distribution is represented by darken the corona (i.e. the darker the corona the increased number of nodes) so the number of nodes should be increased with geometric proportion from the outer parts to the inner ones.

In our deployment model no controlled node distribution is performed. Instead; the direction of the traffic is reversed. Considering the same network model in fig 1, path *B* represents a possible traffic path. Assuming uniform node density $D = N/A$ where

$$D_1 = D_2 =,,,= D_n = N_i/A_i \quad , \ 1 \leq i \leq n \tag{1}$$

$$\text{Intuitively } A_1 < A_2 < ... < A_i \text{ so } N_1 < N_2 < ... < N_i \tag{2}$$

$$N_{i+1}/ N_i = N_i/N_{i-1} = ... = q \text{ and } q > 1 \tag{3}$$

This means that with uniform node distribution, corona *i+1* has number of nodes larger than the adjacent corona *i* (which is nearer to the network center). This is intuitively enforced by the underlying area to be covered by node distribution. So if the traffic is reversed toward the outer corona where data collector nodes would be deployed we could get the same benefit without intentionally perform a controlled non-uniform node distribution.

3.2 Data dissemination strategy

Reversing the direction of the data traffic from inner parts to outer parts and deploying the data collector nodes at the network peripheral are not enough to get the benefit of the sub-optimal balancing of the energy. Without a data dissemination strategy that could take advantage of the previous consideration, the external sink deployment imposes higher communication costs than the centric-sink model. Implicitly communication paths would be longer in terms of hops count. Nodes around the data collector have to carry the data from relatively larger number of far nodes. These nodes are exposed to drain their energy resources and die faster than others forming an early partition. So a data dissemination strategy is required to take advantages of the sub-optimal energy balance that could be obtained by reversing the direction of the traffic. In the following subsection we describe our boundary-peeling data collection protocol that works for both regular (circular, grid) and irregular network topologies. So the network model in fig 1 is relaxed to the general network model described in section 3. The protocol aims to prolong the network lifetime defined as the time until a network partition occurs [2]. Partition could be either a sink-network partition or internal partition where group of nodes form an isolated island.

4   Boundary-Peeling Data Collection Protocol

The proposed data collection protocol consists of three phases: initial, continuous and conditional one. Fig 2, shows the conceptual steps of the data collection protocol.

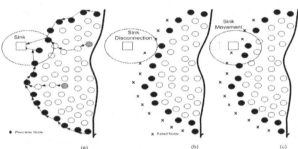

Fig. 2. The boundary-peeling data collection process. (a) The network boundary is recognized and data is directed to the boundary where it travels to the sink. (b) Sink-network partition. (c) Sink movement toward the network center.

4.1 Phase I: Topology recognition

This phase is performed once at the setup time. It consists of three steps. The boundary of the network is recognized at the first step. This step allows each boundary node to know that it is indeed a boundary node. The second step is the computation of the virtual centroid (*v-center*) of the network. This is done offline at one sink (if multiple exist) elected to perform the computation. At the third step the v-center is broadcasted to all nodes. Each node uses the v-center to learn its *allowable forwarding directions* (to be used in phase II). If the center of the network is known in advance as in the case of the circular network model, it could be used directly (so we use the terms *v-center* and network center interchangeably). In reality, networks are neither circular nor regular. Networks with irregular topologies are important class of sensor networks.

We use the *BoundHole* algorithm [19] to define the boundary of the network and allow perimeter nodes to know that they are located at the network boundary. The algorithm was developed to detect the holes within the sensor field. It inherently identifies the outer boundary of the network. Peripheral sinks cooperatively implement the *Boundhole* at the initial phase. Assuming a total of m sinks S_0, S_1,..., S_{m-1}, we divide the boundary into m boundary segments associated with the m sinks. Each segment starts at one sink and ends at the next sink in a counterclockwise direction. So the outer boundary is represented by the following sequence: S_0, p_1, p_2..., p_i, S_1, p_{i+1}, p_{i+2}... , p_k, S_2, p_{k+1}... S_{m-1}... p_j , S_0. where p's are the sink-to-sink nodes located at the outer boundary. Each perimeter node p keeps information about its upstream and downstream p's nodes in addition to counters (registering its hop-count) to each end-sink. A node also creates a pointer to its parent (upstream) node toward the closest end sink. Perimeter nodes would exhibit the heavier load amongst the network nodes. Frequently such nodes would lose their energy and die exposing other interior nodes to be perimeter. The algorithm is supported by a local maintenance process (at level of one-hop neighbors) that allows the replacement of the dead perimeter nodes by fresh ones. Such local maintenance process allows dynamic coping with topological changes affecting the outer boundary and capturing the continuous updates.

Computation of virtual centriod

To compute the *v-center* c, we consider the network as a closed polygon in which a finite set of perimeter nodes makes up a set of virtual vertices, or *v-points*. The *v-center* is defined as the median of this set. The concept of the *v-center* provides only an approximate solution as the network is not really a polygon with known vertices. The *v-points* can be selected using a threshold distance value (h number of hops) as following: once the network perimeter is marked, perimeter nodes that satisfy kh distance (where k =1, 2, 3,.., etc.) to the sink that is elected to perform the computation, report their coordinates. The sink collects the positions of the v-points and determines the coordinates (x_c, y_c) of c as in (4) where A is the virtual network area using the selected v-points [20]. The v-points could also be selected as the ends and middle of each boundary segment. We note that the smaller the value of h, the more precise the *v-center*.

$$x_c= \frac{1}{6A}\sum_{i=0}^{K-1}(x_i + x_{i+1})(x_i y_{i+1} - x_{i+1}y_i), y_c= \frac{1}{6A}\sum_{i=0}^{K-1}(y_i + y_{i+1})(x_i y_{i+1} - x_{i+1}y_i) \qquad (4)$$

Forwarding directions

The *v-center* c computed at the previous step acts as a center of a disperse force that directs the data away from it toward the network peripheral. Each node learns its allowable forwarding directions (out of four basic directions Up, Down, Left, Right) according to its location with respect to c. For each node, the four basic directions are represented by four abstract reference points on its communication circumference (the circle with the node as the center and the communication radius as the radius). The allowable directions are those that direct the data *away-from* the *v-center*. For example, if the location of the *v-center* c with respect to node n_1 is at *up-right* corner of an imaginary rectangle, n_1 excludes both Up and Right directions. n_1 considers the remaining directions (Left and Down) as its allowable forwarding directions. The node alternatively selects its current forwarding direction (the direction where it should transmit its data) out of them.

4.2 Phase II: Data Forwarding

The data forwarding phase is a continuous phase Interior (non-perimeter) nodes forward their data toward the network perimeter. The node selects the current direction of data forwarding and transmits the data to the next forwarder in that direction. The next forwarder is the neighbor closest to the reference point of the chosen direction among those closer to that point than the node itself. The process continues until the data arrives at any perimeter node. Data arrived at the perimeter region travels the perimeter toward the closest sink.

4.3 Phase III: Sink Mobility

The mobility phase is a conditional phase sinks stay stationary at their initial deployment locations until one of the movement conditions is satisfied. Sinks continuously validate the conditions and make the movement decision when required. A condition could be (a) a sink is exposed to a full partition at its current location, (b) the current number of its one-hop neighbors is less than a threshold value and (c) The rate of receiving data at the current location is less than an expected threshold value. The movement pattern is such that the sink moves discrete steps toward the network v-center. So the v-center acts as a center of a pulling force that attracts the moving sinks (in contrast to the "away-from-centroid" dissemination strategy that is assumed for internal data propagation). The moving sink checks the validity of the moving condition in each step and moves to regain its initial situation. In each step, the sink moves a progressive distance equal to its transmission range in the direction of the line joining the current location of the sink to the network center. Once the movement action is completed, the sink informs its current one-hop neighbors of its presence in their transmission region. It sends them an *explore* message and collects their replies. The sink reassigns the role of perimeter neighbors among them and sends a *glue-boundary* message. The perimeter nodes locally glue the boundary within their segment and adjust their hop-count toward the sink. They also modify their pointers to the upstream neighbors toward the closest sink if required. This mechanism maintains the connectivity of sinks to the network boundary.

5   Performance Evaluation

We evaluated our protocol using the wireless sensor network simulator (WSNS) described in [21]. Two categories of experiments are performed to test the behavior of the protocol in both dense and sparse networks. We used network terrain dimensions of 400 m × 400 m with sensor densities ranging from 200 to 400 nodes (randomly distributed). Different radio ranges T_x (ranging from 60m to 90m) are used to maintain the connectivity of the network with the changed densities. Experiments are conducted with different number of sinks deployed dynamically at the network peripheral. In all experiments, the initial energy of sensor nodes is set to 100 units. We assume that each data transmission would consume one unit of energy. For simplicity, the third mobility condition mentioned in the pervious section is not simulated and we stop the simulation when a partition occurs.

A third category of experiments is conducted with centric-sink network model and the performance is compared to our model. In centric-sink model, a single sink is located at the center of the network and the shortest path routing SPR is applied as the underlying routing protocol (this model is considered as an ideal model in [6, 7, 10, 15]). Since nodes are stationary and no underlying MAC protocol is considered, Routing tables of the SPR are determined at the initial phase. We then allow a control packets transmission session every simulation round (to update the routing information and cope with routing failures due to the increase in the number of dead nodes). We assume that each node consumes 0.5 unit of energy to transmit one control message. Results from the simulation are collected at different rounds. Each round is equivalent to 2 data transmission sessions and one control transmission session. To ensure consistency, all the plotted results are the average of 5 runs of each experiment.

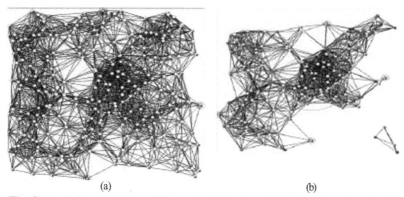

(a)   (b)

Fig. 3. Simulation snapshots at different simulation time, (a) the initial topology, 4 sinks are deployed at the network peripheral (nodes with double circles), perimeter nodes are recognized and each node is marked with small dark rectangular, (b) The network partition (some nodes form an isolated island the simulation stops at this point).

Fig 3, shows snapshots of simulation runs at different simulation rounds. It shows the updates of the network boundary (that we call the peeling phenomenon) in a dense network scenario. It also illustrates the movement of sinks toward the network center (sinks are not necessary moving simultaneously).

For each configuration scenario, the level of each node with respect to the network center is determined Level 1 nodes are those located at distance less than T_x with respect to the network center c. Level 2 nodes are those located at distance larger than T_x and smaller than $2 T_x$ with respect to c and so on.

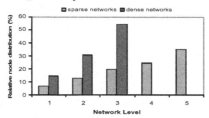

Fig. 4. Relative node distribution

Fig 4 shows relative node densities of different network layers for both dense and sparse network configurations. Layer one is the nearest one to the network center. In dense networks, the nodes are separated over three layers. In sparse networks, nodes are separated over 5 layers. The relative densities are shown in fig 4.

To show how our deployment model with the associated data collection protocol results in load balancing through the network, the *average load* at each network level is measured and depicted in fig 5. The figure shows that our protocol definitely balances the average load at each network layer except the outmost one (network level 5 in sparse settings and 3 in dense settings). The figure also shows that increasing the number of peripheral sinks results in decreasing the average load at the outmost layer. The inner layers would approximately exhibit the same average load. In the case of centric-sink model with shortest path routing, the innermost layer (level 1) is exhibiting the higher load amongst others. The gradual increase in the average load from outmost layer to innermost layer is higher than that exhibited by our model for both sparse and dense networks.

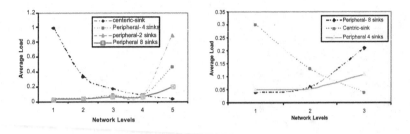

Fig. 5. Comparison of average load at each network level in case of sparse network (Left) and dense network (right).

The *number of dead nodes* is measured at each simulation round and the average rate of dead nodes is depicted in fig 6.

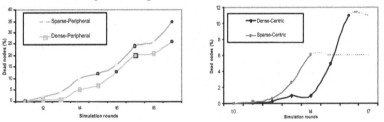

Fig. 6. The rate of dead nodes at different simulation rounds. (Left) Small squares explain the points of movement of one or more sink s and the solid black circle explains the point of partition. (Right) Dotted segments explain the network partition.

The figure shows that our protocol exhibits a higher rate of dead nodes than that exhibited in the case of centric-sink model. Most of these nodes are separated over the network boundary which reduces the negative effect of their early death. Fig. 6 also illustrates the simulation rounds at which the partition occurs. The results are the average results obtained for different network settings where 10 simulation rounds are allowed per each setting. In sparse settings, our model maintains the network connectivity twice the time exhibited using the centric-model. In dense settings, the partition is more delayed than in the case of sparse networks for both models. The partition occurs around the seventh simulation round in case of centric-model while no partition appeared using our model over the 10 simulation rounds.

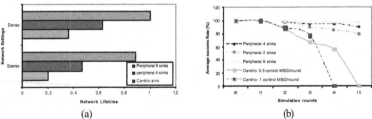

(a)   (b)

Fig. 7. Normalized network lifetime and the average success rate

The normalized network lifetime is shown in fig. (7-a). The figure shows the relative increase in the network lifetime achieved by our protocol. It also shows the effect of increasing the number of peripheral sinks. We also measure the *success rate* of both models in the two network categories. The success rate measures the capability of the live nodes to send their data successfully to the sink. Fig. (7-b) shows the average results obtained for both dense and sparse configurations. Both models demonstrate 100% success rate at the early simulation rounds. Our deployment model with the associated data collection protocol keeps the success rate high during the whole simulation (in both network configurations). The figure shows that at the time a partition occurs the decrease in the success rate is very small. Most of the live nodes are able to send their data successfully to the sink. Fig (3-b) shows an example where only a small group of the live nodes forms an isolated island.

The success rate in case of Centric-sink model heavily depends on the transmission rate of the control messages. Such messages provide freshness of the routing tables. Increasing the rate of control messages increases the success rate yet decreases the network lifetime. This is due to increasing the energy consumption rate at each node which in turn fastens the network partition. When the sink is partitioned from the network the success rate drops to zero even thought a large number of nodes could be alive.

6   Discussions

The proposed deployment model with the associated data collection protocol reduces the negative effect of the peeling phenomenon. The direction of the peeling is from outmost layers to inner ones (external peeling). Since the proposed mobility strategy allows the movement of sinks to follow the direction of the progressive peeling (toward the network center i.e. from outmost to innermost), sinks are allowed to be approximately always connected. Even at the time partition occurs as in fig (3-b) a large part of the network is still connected to one or more sinks and the network is still operable. This raises a question about the coverage percentage that could be tolerated by the application. This feature of our protocol increases the utilization of the network resources. In the case of centric-sink the peeling phenomenon also occurs yet the direction of such peeling is toward the outmost layers. Such internal peeling results in early partition of the sink. The outmost layers could be thought of as *guard layers* that protect the inner layers If the deployment is planned such that the outer layers are assumed to be additional layers to the core network layers, exposing such layers to a peeling would not affect the interested coverage area. The network lifetime would be increasing proportional to the number of nodes in the outer layers. The coverage would be maintained according to the application requirements.

7   Conclusion and Future work

In this paper, we described the boundary-peeling data collection protocol and the associated sink-deployment model. The contributions of our solution includes the following: (i) it combines the benefits of using multiple and mobile sinks without introducing the problems associated with the sink mobility, (ii) It adapts to the topological changes in an efficient and dynamic manner with no overhead to exchange topological updates messages, and (iii) it balances the load among the nodes dynamically. The protocol with the associated deployment model is highly reliable. It is able to maintain the network connectivity as long as the coverage percentage is tolerated by the application. Further investigations of guard regions and the adaptive mobility are left for future work. Our sights are set initially on the development of policy-based sink mobility that adapts to the underlying sensing field conditions.

Reference s

1. Li, J., Mohapatra, P.: An analytical model for the energy hole problem in many-to-one sensor networks In: Proc. of IEEE VT C, Fall 2005.
2. Joongseok Park, Sartaj Sahni: An Online Heuristic for Maximum Lifetime Routing In Wireless Sensor Networks IEEE Trans. Computers 55(8): 1048-1056 (2006).
3. Soro, S., Heinzelman, W.: Prolonging the lifetime of wireless sensor networks via unequal clustering In: Proc. of the19th IEEE International Parallel and Distributed Processing Symposium (IPDPS) , 2005.
4. Li, C. F., Ye, M., Chen, G., Wu, J.: An energy-efficient unequal clustering mechanism for wireless sensor networks In: Proc. of the 2nd IEEE International Conference on Mobile Adhoc and Sensor Systems (MASS), Washington, DC, Nov. 2005.
5. Athanasios, I., Rolim, J.: Fast and Energy Efficient Sensor Data Collection by Multiple Mobile Sinks In: ACM MobiWac 2007, Crete, Greece, October 22, 2007.
6. Luo, J., Panchard, J., Piorkowski, M., Grossglauser, M., Hubaux, J. P.: Mobiroute: Routing towards a mobile sink for improving lifetime in sensor networks. In: IEEE DCOSS, volume LNCS 4026, pages 480--497, 2006.
7. Xiaobing Wu, Guihai Chen, and Sajal Das, : On the Energy Hole Problem of Non-uniform Node Distribution in Wireless Sensor Networks. In: Proceedings of IEEE MASS 2006, October 2006.
8. Akyildiz, I. f., et al.: Wireless sensor networks: a survey. Computer Networks, Vol. 38, March 2002, pp. 393- 422.
9. Al-Karaki, J. N., Kamal, A. E.: Routing techniques in wireless sensor networks: a survey. IEEE Wireless Communications, vol. 11, no. 6, pp. 6–28, Dec. 2004.
10. Olariu, S., Stojmenovic, I.: Design guidelines for maximizing lifetime and avoiding energy holes in sensor networks with uniform distribution and uniform reporting. In: Proc. of IEEE INFOCOM , Apr. 2006.
11. Shi, Y., Thomas Hou, Y.: Approximation Algorithm for Base Station Placement in Wireless Sensor Networks. In Proc. IEEE Communications Society Conference on Sensor and Ad Hoc Communications and Networks (SECON), pp. 512-519, San Diego, CA, June 18-21, 2007.
12. Wang, W., Srinivasan, V., Chua, K.: Using mobile relays to prolong the lifetime of wireless sensor networks," in Proc. Of ACM MobiCom, Aug. 28 - Sept. 2, 2005.
13. Jea, D., Somasundara, A., Srivastava, M.: Multiple controlled mobile elements (data mules) for data collection in sensor networks. In IEEE DCOSS, pages 244--257, 2005.
14. Kinalis, A., Nikoletseas, S.: Scalable data collection protocols for wireless sensor networks with multiple mobile sinks. In 40th ANSS, pages 60–69, March 2007.
15. Luo, J., Hubaux, J. P.: Joint mobility and routing for lifetime elongation in wireless sensor networks In Proc. of IEEE INFOCOM, Mar. 2005.
16. Hashish, S., Karmouch, A.: Topology-based On-Board Data Dissemination Approach for Sensor Network. In ACM MobiWac 2007, Crete, Greece, October 22, 2007
[17] Sadler, B. M.: Fundamentals of energy-constrained sensor network systems. IEEE Aerospace and Electronic Systems Magazine, Tutorial Supplement, 2005.
18. Kansal, A., Rahimi, M., Kaiser, W. J., Srivastava, M. B., Pottie, G. J., Estrin, D.: Controlled mobility for sustainable wireless networks. In Proc., IEEE Sensor and Ad Hoc Communications and Networks (SECON), Santa Clara, CA, Oct. 2004.
19. Fang, Q., Gao, J., Guibas L.: Locating and Bypassing Routing Holes in Sensor Networks. In 23rd Conference of the IEEE Communications Society (InfoCom), 2004.
20. Shamos M. I., Preparata, F.P.: Computational Geometry. Springer-Verlag, Berlin, 1985.
21. http://www.djstein.com/projects/WirelessSensorNetworkSimulator.html

An Architecture for Multimedia Delivery Over Service Specific Overlay Networks

Ibrahim Al-Oqily[1], Ahmed Karmouch[1], and Roch Glitho[2]

[1]School of Information Technology & Engineering (SITE), University Of Ottawa, PO Box 450, Ottawa, ON, K1N 6N5, Canada. {ialoqily, karmouch}@site.uottawa.ca.

[2]Ericsson Canada, 8400 Decarie, Montreal Quebec, Canada, H4P2 N2, roch.glitho@ericsson.com

Abstract. Overlay networks are becoming widely used for delivering content, since they provide effective and reliable services that are not otherwise available. However, overlay management systems face the challenges of increased complexity and heterogeneity due to the numerous entities that are involved in realizing overlay services. We believe that autonomic management is a key solution for dealing with the complexity of overlay management. In this paper, a management architecture for service specific overlay networks is proposed. Overlays are viewed as a dynamic organization for self-management in which self-interested nodes can join or leave according to their goals. The objective of this architecture is to create autonomic overlays that are driven by different levels of policies. Policies are generated at different levels of the autonomic management hierarchy and enforced on the fly. The proposed autonomic management dynamically adapts the behavior of the overlay network to the preferences of the user, network, and service providers. A description of our novel architecture that addresses these challenges is presented.

Keywords— Autonomic Computing, Adaptive SSONs, Network Management, Overlay Networks, Policies.

1 INTRODUCTION

Overlay networks are a virtual topology on top of a physical topology, and they are becoming more popular due to their flexibility and their ability to offer new services. Recent research in this area has focused on designing specific overlay networks to deliver media in heterogeneous environment. For example, SMART [1] was developed in the context of the Ambient Network project [2] in order to optimize media delivery services by moving control and resource allocation to the network itself, since it has more knowledge about topology and network proper-

Please use the following format when citing this chapter:

Al-Oqily, I., Karmouch, A. and Glitho, R., 2008, in IFIP International Federation for Information Processing, Volume 264; Wireless Sensor and Actor Networks II; Ali Miri; (Boston: Springer), pp. 97–112.

ties. SMART creates a Service Specific Overlay Network (SSON) for each media delivery service or group of services. SSON construction uses network side functions, called MediaPorts (MPs). MPs, thereby, provide the flexibility to modify the content and the services, such as caching, adaptation and synchronization [3].

A service delivered to a customer by a SP is usually formed from a composition of different services. Some services are basic in the sense that they cannot be broken down further into component services, and usually act on the underlying resources. Other services are composed of several basic services. Every service consists of an allocation of resource amounts to perform a function. However, with the increasing demands for QoS, service delivery should be efficient, dynamic, and robust. Current manual approaches to service management are costly and consume resources and IT professionals' time, which leads to increased customer dissatisfaction. With the advent of new devices and services the complexity is further increased.

Given the multiple sources of the heterogeneity of networks, users, and applications, constructing SSONs in large distributed networks is challenging. Media content usually requires adaptation before it is consumed by media clients. For example, video frames must be dropped to meet QoS constraints. Other examples are when a client with a PDA requires a scale down for a video, or when content must be cached to be viewed by a mobile user. In addition to adaptation, new users may request to join or leave the overlay, a network node may fail, or a bottleneck may degrade the SSON's performance. Consequently, the overlay must be adapted to overcome these limitations and to satisfy the new requests. It is obvious that with a large number of overlays, the management task becomes harder to achieve using traditional methods. Therefore new solutions are needed to allow SPs to support the required services and to focus on enhancing these services rather than their management.

The concept of autonomic computing (AC) [4] is proposed by IBM to enable systems to manage themselves through the use of self-configuring, self-healing, self-optimizing, and self-protecting solutions. It is a holistic approach to computer systems design and management with the goal to shift the burden of support tasks, such as configuration and maintenance, from IT professionals to technology. Therefore, AC is a key solution for SSON management in heterogeneous and dynamic environments.

Establishing a SSON involves 1) Resource discovery to discover network-side nodes that support the required media processing capabilities. 2) An optimization criterion to decide which nodes should be included in the overlay network. 3) Configuring the selected overlay nodes, and 4) Adapting the overlay to the changing network context, user, or service requirements, and joining and leaving nodes. In AC, each step must be redesigned to support autonomic functions. In other words, in Autonomic Overlays (AO), each step imposes a set of minimum requirements. For example, the resource discovery scheme should be distributed and not rely on a central entity, dynamic to cope with changing network conditions, efficient in terms of response time and message overhead, and accurate in terms of its success rate. The optimization step is mapped into a self-optimization that selects resources based on an optimization criterion (such as delay, bandwidth, etc.) and should yield the cheapest overlay, and/or an overlay with the least number of hops, and/or an overlay that is load-balanced, and/or a low latency overlay network, and/or a high bandwidth overlay network. The configuration of the selected overlay nodes in a given SSON is mapped into a self-configuration and self-adaptation. Self-configuring SSONs dynamically configure themselves on the fly. Thus they can adapt their overlay nodes immediately to the joining and leaving nodes and to the changes in the network environment. Self-adapting SSONs self-tune their constituent resources dynamically to provide uninterrupted service. Our goals are to automate overlay management in a dynamic manner that preserves the flexibility and benefits that overlays provide, to extend overlay nodes to become autonomic, to define the inter-node autonomic behavior between overlay nodes, and to define the global autonomic behavior between SSONs.

This paper proposes a novel management architecture for overlay networks. Our contributions are twofold. First we introduce the concept of Autonomic Overlays (AO), in which SSONs and their constituent overlay nodes are made autonomic and thus become able to self-manage. Second, autonomic entities are driven by policies that are generated dynamically from the context information of the user, network, and service providers. This ensures that the creation, optimization, adaptation, and termination of overlays are controlled by policies, and thus the behaviors of the overlays are tailored to their specific needs.

The paper is organized as follows. Section 2 discusses the related work. Section 3 introduces the proposed autonomic overlay architecture. Section 4 discusses the experimental evaluation. Finally, in Section 5, we draw our conclusion and suggest future work.

2   RELATED WORK

The proposed autonomic overlays draw upon IBM's vision and blueprint [7]. The work presented in this paper is concerned with all possible phases of the service delivery in SSONs – from the instant of requesting a service to the instant of terminating it. Thus we present an integral approach to SPs wishing to deliver services over their infrastructure.

IBM identified the complexity of current computing systems as a major barrier to its growth [5]. The solution to this problem lies in more intelligent systems called autonomic computing (AC). AC simplifies and automates many system management tasks traditionally carried out manually. Systems that manage themselves are able to adapt to changes in their environment in accordance with business objectives. The result is a great savings in management costs and IT professionals' time. This will free the latter to focus on improving their offered service rather than managing them manually. Some of the main scientific and engineering challenges that collectively make up the grand challenge of autonomic computing were outlined in [6]. Also, a set of characteristics required by AC were identified and explained in [7].

According to the IBM vision [4], an AC system is a system that knows itself and its environment, configures and reconfigures itself under varying and unpredictable conditions, heals itself, provides self-protection, and keeps its complexity hidden. Although the IBM vision is a holistic approach to designing computer systems, much of the research in this field focuses on a few specific aspects of this vision.

Autonomic communications were proposed in [8]. It has a similar concept to IBM's autonomic computing. The difference is that, in the former, the focus is on individual elements of the network, how their behavior is learned and altered, and how they interact with other elements. Our work focuses on service specific overlay networks; thus, the interaction between the network and computing entities is based on a service request/offer concept in which each entity is responsible for its internal state and resources. An entity may offer a service to other entities. The offering entity responds to a request based on its willingness to provide a service in its current state. A generic architecture for autonomic service delivery was proposed in [9]. It defines a resource management model based on virtualization. However, it is service-independent, and is unlikely to achieve the specific QoS requirements for each service dynamically without human intervention. A model for dynamic fault tolerance technique selection for grid work flow, that allows the system to configure its fault tolerance mechanism, was developed in [10].

Policy-based management for computer systems has also been studied. Pattern classification and clustering techniques that support online decision making and incremental learning in autonomic systems were proposed in [11]. The use of policies to configure autonomic elements to enforce the required behavior in an Apache web server was presented in [12]. A set of UML-based models were developed and used in [13] to specify autonomic properties and to deploy policies as an executing system based on composition and model modification. A policy-driven model based on multi-agent systems was also proposed in [14]. In their model, Web services are represented as agents and agent behavior is controlled using high level policies. A mapping of biological systems to PBMS was introduced in [15]. Their system is hierarchical and relies on mechanisms for organism regulation, which supports self-management at different levels of the hierarchy. Humans in an organization thus specify policy at a level of abstraction that reflects their specific needs. The difference between our work and all these approaches is that the above approaches consider a particular service to which their design is appropriate. In addition, policy generation is not a fully automatic process and human intervention is still needed.

Projects such as Service Clouds [16], Autonomia [17], GridKit [18], Auto-Mate [19], and Unity [20] are using the autonomic concept in different ways. Service Clouds provides an infrastructure for composing autonomic communication services. It combines adaptive middleware functionality with an overlay network to support dynamic service reconfiguration. Autonomia provides dynamically programmable control and management to support the development and deployment of smart applications; primarily, it achieves the self-healing property for failed entities. GridKit proposes a middleware that offers a consistent programming model across different communication types. AutoMate enables the development of autonomic Grid applications by investigating programming models, frameworks, and middleware services that support autonomic elements. Finally, Unity designs both the behavior of individual autonomic elements and the relationships that are formed among them in order to create computing systems that manage themselves. A detailed survey on autonomic computing is available in [21].

3   AUTONOMIC OVERLAYS

To tackle the complexity of overlay management, each SSON is managed by an SSON Autonomic Manager (SSON-AM) that dictates the service performance parameters. This ensures the self.* functions of the service. In addition to this, overlay nodes are made autonomic to self-manage their internal behavior and their interactions with other overlay nodes. In order to ensure system wide performance, System Autonomic Managers (SAM) manages the different SSON managers by providing them with high level directives and goals. The following sections detail the different aspects of our architecture.

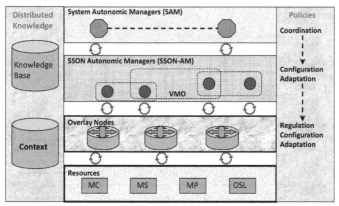

Fig. 1. Autonomic overlays architecture

3.1 Architecture Overview

The set of components that makes up our architecture is shown in Fig. 1. The lowest layer contains the system resources that are needed for multimedia delivery sessions. In particular, the Overlay Support Layer (OSL) receives packets from the network, sends them to the network, and forwards them on to the overlay. Overlay nodes implement a sink (MediaClient, or MC), a source (MediaServer, or MS), or a MediaPort (MP) in any combination. MPs are special network side components that provide valuable functions to media sessions; these functions include, but are not limited to, special routing capabilities, caching, and adaptation. These managed resources can be hardware or software and may have their own self-managing attributes.

The next layer contains the overlay nodes. Overlay nodes are physical Ambient Network nodes that have the necessary capabilities to become part of the SSON. They consist of a control plan and a user plan. The control plan is responsible for the creation, routing, adaptation, and termination of SSONs, while the user plan contains a set of managed resources. The self-management functions of overlay nodes are located in the control plan. The Ambient manageability interfaces are used by the self-managing functions to access and control the managed resources. The rest of the layers automate the overlays' management in the system using their autonomic managers. SSON-AMs and SAMs may have one or more autonomic managers, e.g. for self-configuring and self-optimizing. Each SSON is managed by an SSON-AM that is responsible for delivering the self-management functions to the SSON. The SAMs are responsible for delivering system wide management functions; thus, they directly manage the SSON-AMs. The management interactions are expressed through policies at different levels. All of these components are backed up with a distributed knowledge. The following sections describe each component in detail.

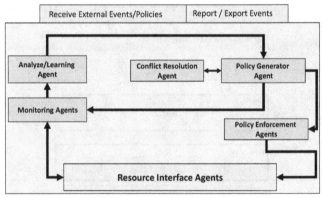

Fig. 2. The set of components that makes up the intelligent control loop

3.2 Autonomic Elements

1) Overlay Nodes Autonomic Manager(ONAM): Each overlay node contains a control loop similar to the IBM control loop [5] as shown in Fig. 2. The

Autonomic Manager (AM) collects the details it needs from its managed resources, analyzes those details to decide what actions need to change, generates the policies that reflects the required change, and enforces these policies at the correct resources. As shown in the figure, the ONAM consist of the following:

- Monitoring Agents (MAs): collects information from the overlay node resources, such as packet loss, delay jitter, and throughput. It also correlates the collected data according to the installed policies and reports any violation to the Analyze/Learning Agent (ALA). For example, an MA for a caching MP collects information about the MP's available capacity, and whenever the available capacity reaches 10% it reports to the ALA. Another example is the MA for a routing MP that relays data packets between overlay nodes: its MA collects information about the throughput and reports to the ALA whenever the throughput reaches a high value. These collected data will be used to decide the correct actions that must be taken to keep the overlay node performance within its defined goals. The MAs interact with the Resource Interface Agents (RIAs) to monitor the overlay node resources availability and to collect data about the desired metrics. They also receive policies regarding the metrics that they should monitor as well as the frequency in which they report to the ALA.

- Analyze/Learning Agent (ALA): observes the data received from the MAs and checks to see whether a certain policy with which its overlay node is abided is not being met. It correlates the observed metrics with respect to the contexts, and performs analysis based on the statistical information. In the case that one of policies is violated, it sends a change request to the Policy Generator (PG). This component is an objective of future work.

- Policy Generator (PG): The difference between this control loop and the IBMs' control loop lies in the use of a PG instead of a Plan component. The Plan function –according to IBM [5] – is to select or create a procedure that reflects the desired change based on the received change request from the Analyze Agent. This is not sufficient in our case, where each overlay node receives high level policies and it is up to the overlay node to decide how to enforce these policies based on its available resources. Therefore, we envisioned a PG instead. The PG reacts to the change request in the same way as in the Plan component, although it also generates different types of policies in response to the received high level policies. For example, based on the goal policies received by the overlay node, the policy generator generates the tuning polices and passes them to the MAs (more about this in Section 3.4). Upon generating new policies, the policy generator consults a Conflict Resolution Agent (CRA) that ensures the consistency of the new generated policies with those that already exist. Generally, we divide conflicts into two types: static conflicts and dynamic conflicts. In our model, a static conflict is a conflict that can be detected at the time of generating a new policy, while a dynamic conflict is one that occurs at run time.

- Policy Enforcement Agent (PEA): The PG generates suitable policies to correct the situation in response to a change request, and passes these policies to the

PEA. The PEA then uses the suitable RIA to enforce them. This includes mapping the actions into executable elements by forwarding them to the suitable RIA responsible for performing the actual adjustments of resources and parameters. The enforced policies are then stored in the Knowledge Base (KB).

- Resource Interface Agents (RIAs): These implement the desired interfaces to the overlay node resources. The MAs interacts with them to monitor the availability of overlay node resources and the desired metrics in its surrounding environment. Each resource type has its own RIA that translates the policy actions into an adjustment of configuration parameters that implements the policy action.
- Each overlay node has a set of interfaces to receive and export events and policies to other overlay nodes. These interfaces are essential to enable multiple overlay nodes to cooperate to achieve their goals. In particular, these interfaces are used by the SSON-AM to interact with the overlay nodes that agreed to participate in the SSON. The SSON-AM sends the system policies to the overlay nodes through these interfaces, through which it also receives reports on their current status.

2) SSON Autonomic Managers (SSON-AM): SSON-AMs implement the intelligent control loop in much the same way as ONAMs. They automate the task of creating, adapting, configuring, and terminating SSONs. They work directly with the ONAM through their management interfaces. They perform different self-management functions, such as self-configuring, self-optimizing, and self-adapting. Therefore, they have different control loops. Typically, they perform the following tasks:

- Self-configuration: SSON-AMs generate configuration policies in response to the received system policies. They use these policies to configure overlay nodes that are participating in a given SSON.
- Self-optimization: during SSON construction, SSON-AMs discover the overlay nodes required to set up a routing path for the multimedia session. Therefore, they are responsible for optimizing the service path to meet the required QoS metrics induced from high level policies as well as the context of the service.
- Self-Adaptation: SSON-AMs monitor the QoS metrics for the multimedia session and keep adapting the service path to the changing conditions of the network, service, and user preferences. They also monitor the participating overlay nodes and find alternatives in case one of the overlay nodes is not abiding to the required performance metrics.

SSON-AMs receive goal policies from the SAMs to decide the types of actions that should be taken for their managed resources. A SSON-AM can manage one or more overlay nodes directly to achieve its goals. Therefore, the overlay nodes of a given SSON are viewed as its managed resources. In addition, they expose manageability interfaces to other autonomic managers, thus allowing SAMs to interact with them in much the same way that they interact with the ONAMs. See Fig. 3.

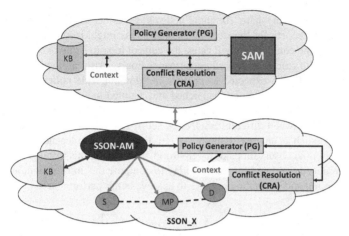

Fig. 3. The lower part represents an SSON that consists of a Source (S), a Destination (D), and a MediaPort (MP). The SSON is managed by a SSON-AM. It has its own Knowledge Base (KB). The upper part represents a SAM and its components.

3) System Autonomic Managers (SAM): A single SSON-AM alone is only able to achieve self-management functions for the SSON that it manages. If a large number of SSONs in a given network with their autonomic managers is considered, it is observable that these SSONs are not really isolated. On the one hand, each overlay node can be part of many SSONs if it offers more than one service or if it has enough resources to serve more than one session. On the other hand, the SSONs' service paths may overlap, resulting in two or more SSONs sharing the same physical or logical link. For example, consider two SSONs sharing the same routing MP with the same goal to maximize throughput. This will lead to a competition between autonomic managers that are expected to provide the best achievable performance. Therefore, and in order to achieve a system wide autonomic behavior, the SSON-AMs need to coordinate their self-managing functions. Typically this is achieved using SAMs.

SAMs can manage one or more SSON-AMs. They pass the system high level policies, such as load balancing policies, to the SSON-AMs. Moreover, whenever they find shared goals between two different SSON-AMs, they inform them to avoid conflicting actions. The involved autonomic managers then contact each other and create a Virtual Management Overlay (VMO) among themselves as illustrated in Fig. 4. They use this VMO to coordinate their management actions before they are passed to their overlay nodes.

Sharing goals is not the only reason to create VMOs; SSONs sharing common links as well as SSONs that belong to the same policy domain (same service class, ISP, etc.) may also create VMOs among themselves to coordinate their actions. Moreover, SSONs that share common nodes/links affect each other's performance, as they compete for the shared resources. This might result in a degraded performance as the competition will cause the control loop to be invoked frequently in an attempt to reach the desired performance goals. Also, all the SSONs in a given domain (ISP) are expected to achieve the domain wide policies together. VMOs allow these policies to be dispatched and adapted to each SSON in a way that achieves the desired goals. Moreover, VMOs also allow the sharing of control and information between different SSONs. A set of SSONs that are co-located in given vicinity (such as an area, domain, AS, etc.) usually has independent rout decisions based on its observations of its environment. Sharing this information will result in a reduced overhead for each overlay to compute this information, and allows for adapting and generating policies to achieve better performance.

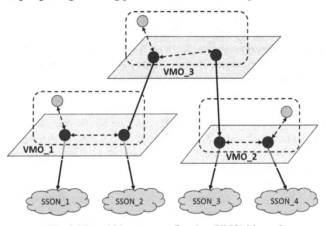

Fig. 4. Virtual Management Overlay (VMO) hierarchy

3.3 Distributed Knowledge

Each autonomic manager obtains and generates information. This information is stored in a shared Knowledge Base (KB) (see Fig. 3). The shared knowledge contains data such as SSON topology, media type descriptions, the set of policies that are active, and the goal policies received from higher level autonomic managers. The shared knowledge also contains the monitored metrics and their respective values. When VMOs are created, each autonomic manager can obtain two types of information from its VMO peers. The first is related to the coordination actions and the second is related to the common metrics in which each autonomic manager is interested. Therefore, knowledge evolves over time; the autonomic manager's functions add new knowledge as a result of executing their actions, obsolete knowledge is deleted or stored in log files, and autonomic managers in VMOs exchange and share knowledge. Also, goal policies are passed from high level autonomic managers to their managed autonomic managers. The context information of the network, users, and services is also used primarily to aid in generating suitable policies at each level of autonomic managers.

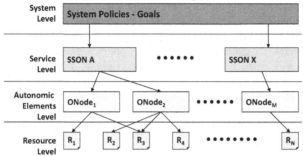

Fig. 5. Policy Levels

3.4 Policies

The use of policies offers an appropriately flexible, portable and customizable management solution that allows network entities to be configured on the fly. Usually, network administrators define a set of rules to control the behavior of network entities. These rules can be translated into component-specific policies that are stored in a policy repository and can be retrieved and enforced as needed. Policies represent a suitable and efficient means of managing overlays. However, the proposed architecture leverages the management task to the overlays and their logical elements, thus providing the directives on which an autonomic system can rely to meet its requirements. Policies in our autonomic architecture are generated dynamically, thereby achieving an automation level that requires no human interaction. In the following we will highlight the different types of policies specific to autonomic overlays. These policy types are being generated at different levels of the system.

Configuration policies: are policies which can be used to specify the configuration of a component or a set of components. The SSON-AMs generate the configuration polices for the service path that meets the SSON's QoS requirements. The ONAMs generate the specific resource configuration policies that, when enforced all together, achieve the SSON QoS metrics. The user, service, and network context are used by these autonomic managers to generate configuration policies.

Adaptation policies: are policies that can be used to adapt the SSON to changing conditions. They are generated in response to a trigger fired by a change in the user, service, or network context. SSON-AMs receive these triggers either from the

SAMs or from the ONAMs, while the ONAMs receive these triggers either from the SSON-AMs or from their internal resources. Whenever a change that violates the installed policies occurs, an adaptation trigger is fired. The autonomic manager that first detects this change tries to solve the problem by generating the suitable adaptation policies; if it does not succeed, it informs the higher level autonomic manager.

Coordination policies: are policies which can be used to coordinate the actions of two or more SSON-AMs. They are generated by the SAMs to govern the behavior of SSON managers that have conflicting goals to avoid race conditions.

Regulation policies: are generated by the overlay nodes themselves to control the MAs' behavior with respect to their goals. For example, a MA that measures throughput has a policy to report throughput < 70%. Another regulation policy can be installed to replace this policy and report throughput < 90%. The second regulation policy can be generated in response to an adaptation policy that requires throughput to be at least 90%. The MAs therefore are made more active to contribute to achieving the required tasks.

Figure 5 shows how these policies are related to our autonomic architecture. At the highest level the SAMs define the set of system polices. These policies represent the system wide goals and do not describe either the particular devices that will be used to achieve the system goals or the specific configurations for these devices. SAMs pass these policies to the SSON-AMs. SSON-AMs refine the system policies and generate service specific policies. They do so by adding further details to the system policies. These details are induced from the system policies as well as from the context information of the users, the network, and the service. At this level, the goals of the SSON under discussion, such as the permitted QoS metrics, are defined. These goals are still device independent policies. The set of service polices is then passed to the ONAMs. These autonomic managers further refine the received policies and generate the overlay node polices and their respective resource specific policies. Overlay node policies represent the goals that this overlay node is expected to achieve, while resource specific policies represents the actual actions that the resources of the overlay node has to do to achieve the overlay node goals. This separation of policies allows each autonomic element to focus on its goals and how to achieve them using its current resources while contributing at the same time to the overall system performance. By decoupling the functionality of adapting overlay node resources policies from the task of mapping system objectives and abstract users' requirements, the policy separation offers users and IT professionals the freedom to specify and dynamically change their requirements. The hierarchical policy model is used to facilitate the mapping of higher level system policies into overlay node objectives. Given sets of user, service and network context and constraints, as well as sets of possible actions to be taken, decisions for policy customizations are taken at run time based on values obtained from MAs to best utilize the available overlay node resources.

In addition to generating policies from high level goals, the policy generator located in each autonomic manager serves as a Policy Decision Point (PDP) for the low level autonomic manager. For example, the SSON-AM serves as a PDP for the ONAM. Whenever an ONAM detects that one of the configuration policies has been violated, it tries to solve the problem locally. If it is unable to do so, it consults the SSON-AM to which the overly node is providing a service. The SSON-AM then tries to solve the problem by either relaxing the goals of the services or by finding an alternative overlay node that is able to achieve the SSON's goals. The SSON-AM then informs the ONAM of its decision, and may also consult its designated SAM to acquire decisions on situations that it cannot handle locally. The autonomic manager acting as a PDP decides which policies, if any configuration or adaptation policies have been violated, were most important and what actions to take. It uses information about the installed policies and the current context of the user, network, and service.

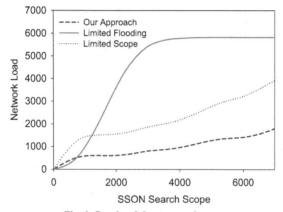

Fig. 6. Overhead due to search messages.

4   EXPERIMENTAL EVALUATION

We used a discrete event simulator to evaluate the performance of the architecture. The measurements of search overhead, and success rate were tested in a large-scale network.

Our first concern was to compare the performance of the architecture in building SSONs with limited-flooding and limited-scope approaches. Limited-flooding has been predominantly used to discover services in environments such as ad hoc and pervasive networks. In a limited-flooding protocol, a service request is broadcast to all direct neighbors of the requesting node. Close neighbors send it on to their neighbors; the propagation is controlled by a TTL value that indicates how far the query should be sent from the requesting node. In a limited-scope approach, the service request is sent to nodes that bring it progressively closer to the destination and located within a scope angle.

The topology used had 2000 nodes in a 1000 X 1000 node two-dimensional overlay space. The bandwidth assigned to each node was randomly selected between 128 and 512 kbits/s. The links propagation delay was fixed at 1 ms. To follow a flash crowd characteristic, all nodes issued their queries at a random point during the first 30 seconds, with the simulation lasting for another 1000 seconds. We ran the simulation a number of times with different search scope values (relative value, similar to TTL except it measures how far the query travels in the network in terms of network distance which is a relatively stable characteristic).

4.1 Network Load

This quantifies the cost of each approach. That is the total number of messages used to construct an SSON. Fig. 6 shows that limited-flooding has the worst performance: it produces a greater number of search messages, except in searches with small scope values. But as the search scope increases, the number of messages in limited-flooding and limited-scope is at least two times higher than the number of messages in our approach.

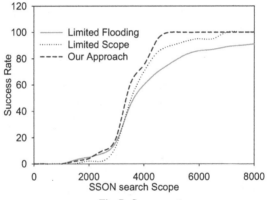

Fig. 7. Success rate.

4.2 Success Rate

Success rate measures the accuracy of each approach. Success rate is defined as the number of requests that receive positive responses, divided by the total number of requests. Fig. 7 shows that our approach results in a higher success rate. Limited-flooding reach the 100% success rate after a 4000 search scope value. While limited-scope approach attains it after a 6500 search scope value. However, our approach reaches the 100% success rate earlier. We believe that this is due to the huge network load generated by limited-flooding. For large search scope values, limited-flooding generates a large number of messages and receives a large number of reply messages. As a consequence, messages are dropped or lost due to collisions.

5   CONCLUSION AND FUTURE WORK

This work-in-progress provides a complete integrated architecture for autonomic SSONs management. It shows how it can be useful to avoid the complexity of existing service management systems. The road towards fully autonomic system architecture is still long; however, this paper presents an autonomic overlay architecture that represents the basic building blocks needed by autonomic systems. The creation, provision, management and termination of SSONs are automated dynamically based on the context information available from the user, service, and the network provider. The required knowledge capability, reasoning capability, and the different autonomic manager components capabilities are being studied. The PG and the CRA components are being investigated in terms of their requirements and implementation to generate different types of policies. The dual goals of these policies are to drive the autonomic system and dynamically manage SSONs.

Acknowledgments

This work was partly supported by a Strategic Research Grant from the Natural Sciences and Engineering Research Council of Canada.

REFERENCES

1. S. Schmid, F. Hartung, M. Kampmann, S. Herborn, and J. Rey, "SMART: Intelligent Multimedia Routing and Adaptation based on Service Specific Overlay Networks," In Proc. of Eurescom Summit 2005, Heidelberg, Germany. pp. 69-77, 2005.
2. N. Niebert, A. Schieder, H. Abramowicz, G. Malmgren, J. Sachs, U. Horn, C. Prehofer, and H. Karl, "Ambient Networks -- An Architecture for Communication Networks Beyond 3G," IEEE Wireless Communications (Special Issue on 4G Mobile Communications -- Towards Open Wireless Architecture), 2004.
3. W. T. Ooi, and R. van Renesse, "The design and implementation of programmable media gateways", In Proc.NOSSDAV'00, Chapel Hill, NC, Jun. 2000.
4. J.Kephart, and D.Chess, "The vision of autonomic computing," IEEE Computer Mag.. Vol. 36, No.1, PP.41-50. Jan. 2003.

5. IBM Corporation, "An architectural blueprint for autonomic computing," White Paper, Jun. 2006.
6. IBM Corporation,"Autonomic computing - a manifesto," http://www.research.ibm.com/autonomic /manifesto/, Oct. 2001.
7. J.Kephart , "Research Challenges of Autonomic Computing," Proc. of the 27th int. conf. on Software Engineering (ICSE'05), St. Louis, Missouri, P. 15 – 22, May 15–21, 2005.
8. A. C. Forum, "Autonomic communication." http:// www.autonomic-communication.org.
9. R.Farha and A.Leon-Garcia, "Blueprint for an Autonomic Service Architecture" 2006
10. J.Nichols, H.Demirkan, and M.Goul, "Autonomic Workflow Execution in the Grid," IEEE Tran. on Systems, Man, and Cybernetics—Part C: Applications And Review, VOL. 36, NO. 3, MAY 2006.
11. E.Kasten and P.McKinley, "MESO: Supporting Online Decision Making in Autonomic Computing Systems," IEEE Trans. on Knowledge And Data Engineering, VOL. 19, NO. 4, April 2007.
12. R.Bahati, M.Bauer, E.Vieira , O. Baek, and C.Ahn, "Using Policies to Drive Autonomic Management," Proc. of the Int. Sym.. on a World of Wireless, Mobile and Multimedia Networks (WoWMoM'06), 2006.
13. J. Pena, M.Hinchey, R. Sterritt A.Ruiz-Cortes and M Resinas, "A Model-Driven Architecture Approach for Modeling, Specifying and Deploying Policies in Autonomous and Autonomic Systems," Proc.of the 2nd IEEE Int. Symposium on Dependable, Autonomic and Secure Computing (DASC'06), 2006
14. F.Zhang, J.Gao, B.Liao, "Policy-Driven Model for Autonomic Management of Web Services Using MAS," Proc. of the 5th Int. Conf. on Machine Learning and Cybernetics, Dalian, 13-16 Aug. 2006.
15. S.Balasubramaniam, K.Barrett, W.Donnelly, S.Meer and J.Strassner, " Bio-inspired Policy Based Management (bioPBM) for Autonomic Communications Systems," Proc. of the 7th IEEE Int. Work. on Policies for Distributed Systems and Networks (POLICY'06), 2006
16. P.McKinley, F.Samimi, J.Shapiro, and C.Tang, "Service Clouds: A Distributed Infrastructure for Constructing Autonomic Communication Services," Proc. of the 2nd IEEE Int. Symposium on Dependable, Autonomic and Secure Computing (DASC'06), 2006.
17. D. Xiangdong, S.Hariri, L.Xue, H.Chen, M.Zhang, S.Pavuluri, and S.Rao, "Autonomia: an autonomic computing environment," in Proc. of the IEEE Int. Performance, Computing, and Communications Conf., pp. 61–68, Apr. 2003.
18. P.Grace, G.Coulson, G.Blair, L.Mathy, W.Yeung, W.Cai, D. Duce, and C. Cooper, "GRIDKIT: pluggable overlay networks for grid computing," in Proc. of the distributed objects and applications conf. (DOA'04), Cyprus, p1463-81, October 2004.
19. M.Parashar, H.Liu, Z.Li, V.Matossian, C.Schmidt, G.Zhang, and S.Hariri, "AutoMate: enabling autonomic applications on the grid," Cluster Computing, Vol. 9, No. 6, PP.161–174, 2006.
20. D.Chess, A.Segal, I.Whalley, and S.White, "Unity: Experiences with a Prototype Autonomic Computing System," Proc. of the Int. Conf. on Autonomic Computing (ICAC'04), 2004.
21. S.Dobson, S.Denazis, A.Fernández, D.Gaïti, E.Gelenbe, F.Massacci, P.Nixon, F.Saffre, N.Schmidt, F.Zambonelli and A.Fernández, "A survey of autonomic communications," ACM Transactions on. Autonomous and Adaptive Systems, Vol. 1, No. 2, P. 223-259, Dec.2006.

A Security Protocol for Wireless Sensor Networks

Chang N. Zhang [1], Qian Yu [1], Xun Huang [1], Cungang Yang [2]

[1] Department of Computer Science, University of Regina
Regina, SK, S4S 0A2, Canada
{zhang, yu209, huang24x}@cs.uregina.ca

[2] Department of Electrical Engineering, Ryerson University
Toronto, ON, M5B 2K3, Canada
cungang@ee.ryerson.ca

Abstract. In this paper, we present an security protocol for Wireless Sensor Networks (WSNs). It is based on the forward and backward property of RC4 states and achieves data confidentiality, data authentication, data integrity, and data freshness with low overhead and simple operation. Furthermore, an RC4-based hash function for the generation of Message Authentication Code (MAC) is presented. The proposed protocol is an ideal solution for wireless sensor networks and other resource-constrained devices where the communication nodes have limited power resources and computational capabilities, and can be widely used in the applications of one-to-one communications as well as broadcasting and multicasting.

Keywords: RC4, Security Protocol, Hash Function, Backward Property of RC4 States, Wireless Sensor Networks, Resource-constrained Communications

1 Introduction

Wireless sensor networks (WSNs), radio frequency identification devices (RFIDs), and some of other resource-constrained applications are being deployed today and will soon become an important part of our infrastructure. Security is a critical factor of these applications due to their impact on privacy, trust and control [1].

Size and cost constraints on resource-constrained devices result in corresponding constraints on resources such as processor capacity, memory, power, and bandwidth. For example, current wireless sensor devices use simple, battery powered 4-bit or 8-bit processors and small amount of memory to perform a few bit-wise and simple arithmetic operations. However, traditional cryptographic algorithms are too complex and power hungry. In order to provide secure communications to resource-constrained devices, lightweight should be the first concern.

A. Perrig et al presented a set of security protocols named SPINS [2] for wireless sensor networks. SPINS has two security building blocks: SNEP provides data

Please use the following format when citing this chapter:

Zhang, C.N., et al., 2008, in IFIP International Federation for Information Processing, Volume 264; Wireless Sensor and Actor Networks II; Ali Miri; (Boston: Springer), pp. 113–124.

confidentiality, two-party data authentication, as well as data freshness, and μTESLA provides broadcast authentication. SPINS is a typical security protocol for wireless sensor networks, especially it avoids asymmetric cryptography to achieve broadcast authentication. SPINS provides data confidentiality by using RC5 block cipher. According to the analysis, stream ciphers are almost always faster and use far less code than do block ciphers, and the operations provided by most ultra-low devices are limited to bit-wise logic operations and simple arithmetic operations [3, 4]. Hence block cipher is too heavy and stream cipher should be considered.

State Based Key Hop (SBKH) protocol [5] is an RC4-based one-to-one security protocol in which two communication nodes share the common knowledge of the RC4 states. SBKH introduces offset to avoid RC4 weak key issue and continually updates and reuses RC4 state for the whole communication process. But SBKH suffers from fake acknowledgement attacks and does not support multicasting and broadcasting due to the strong resynchronization requirement.

In this paper, we present an RC4-based security protocol for communications on wireless sensor networks. Firstly, we indicate that the RC4 state has the forward and backward property. Secondly, we present a data transmission scheme based on RC4 and its backward property, and claim that it greatly reduces the computation overhead and eliminates the strong resynchronization requirement. Thirdly, an RC4-based hash function is presented to generate MAC for data authentication and data integrity. Our simulation and analysis shows that the proposed protocol does not reduce the security strength provided by original RC4 and it works lightly with less implementation complexity and cost. By using offset, the security performance is improved.

2   RC4 and its Backward Property

Stream ciphers and block ciphers are two classes of encryption algorithms. Stream ciphers encrypt individual byte or bit of plaintext one at a time, using a simple time-dependent encryption transformation. Block ciphers simultaneously encrypt groups of characters of a plaintext message using a fixed encryption transformation [3]. Stream ciphers and block ciphers have their respective characteristics, but the stream ciphers are almost always faster and use far less code than do block ciphers [3, 4]. RC4 is probably the most widely used stream cipher nowadays due to its simplicity and high efficiency. RC4 is a variable key-size stream cipher based on a 256-byte secret internal state and two one-byte indexes. The data are encrypted by XORing data with the key stream which is generated by RC4 from a base key. RC4 consists of two parts which are key-scheduling algorithm (KSA) and pseudo-random generation algorithm (PRGA). For a given base key, KSA generates an initial permutation state denoted by S_0. PRGA is a repeated loop procedure and each loop generates a one-byte pseudo-random output as the stream key. That is: at each loop, a one-byte stream key is generated and it is XORed with one-byte of the plaintext, in the meantime a new 256-byte permutation state S as well as two one-byte indices i and j are updated, which defined by $(S_{k+1}, i_{k+1}, j_{k+1}) = \text{PRGA}(S_k, i_k, j_k)$ where i_{k+1} and j_{k+1} are the indices and S_{k+1} is the state updated from i_k, j_k, and S_k by applying one loop of PRGA.

On the issue of the security strength of the RC4, a number of papers have been published to analyze the possible methods of attacking RC4, but none is practical with a reasonable key length, such as 128 bits [4]. WEP is an algorithm to secure IEEE 802.11 wireless network and it requires the use of RC4. Beginning in 2001, several serious weakness (including [6, 7]) were reported and they demonstrate that WEP protocol is vulnerable in a number of areas. In essence, the problem is not with RC4 itself but the way in which keys are generated for use as input to RC4.

In this paper, we present an RC4-based security protocol which maintains the simplicity and efficiency of the RC4, and eliminates its limitations (e.g. it allows delayed data packet). It is based on the following property of RC4 states.

Theorem 1: If $(S^*, i^*, j^*) = \text{PRGA}^k(S, i, j)$, then it has $(S, i, j) = \text{IPRGA}^K(S^*, i^*, j^*)$ where PRGA^k denotes applying PRGA by k loops (same for IPRGA^k) and IPRGA is the reverse algorithm of PRGA (without generating a stream key).

Theorem 1 indicates that any former RC4 state can be recovered from later RC4 state by applying IPRGA corresponding loops. The operation codes of PRGA and IPRGA are depicted in Figure 1 and the proof of the Theorem 1 is provided in Appendix. Thereby, we can conclude that any RC4 state can be forward to a new RC4 state by PRGA and backward to a previous RC4 state by IPRGA. For example, it is easy to generate a previous RC4 state from current RC4 state.

$PRGA(S)$
$Generation\ loop:$
$i \leftarrow (i+1) \bmod 256$
$j \leftarrow (j + S[i]) \bmod 256$
$S[i] \leftrightarrow S[j]$
$Output \quad z \leftarrow S[(S[i] + S[j]) \bmod 256]$

$IPRGA\ (S, i, j)$
$Generation\ loop:$
$S[i] \leftrightarrow S[j]$
$j \leftarrow (j - S[i] + 256) \bmod 256$
$i \leftarrow (i - 1) \bmod 256$

Figure 1: The operation codes of PRGA and IPRGA

3   RC4-based Security Protocol

In this section, an RC4-based security protocol for wireless sensor networks (WSNs) is proposed.

3.1   Terminologies and Notations

The terminologies and notations used in this section are described below:

RC4 State: An RC4 state is a permutation state with 256 state elements and 2 indices (i and j) denoted by (S, i, j). Each state element and each index is of 8 bits in length, making the overall RC4 state to be of 258 bytes in total.

Corresponding RC4 State (CRS): A certain RC4 state generates a certain stream key by RC4, so the same RC4 state should be used for both in encryption and decryption for a given message. In the proposed protocol, each communication node keeps an RC4 state called corresponding RC4 state (CRS). The CRS is to be updated after the encryption or decryption of each fixed length data packet.

Offset (O): Offset is an integer which indicates the number of loops to applying PRGA. The offset loops of PRGA takes place after obtaining the initial permutation state S_0 from KSA, but before encrypting or decrypting message. It carries out only after the base key change takes place. We also use offset to name the process of applying the offset loops of PRGA. The purpose of using offset is to discard the offset number of bytes from the beginning of the stream to avoid major statistical bias in the distribution of the initial bytes of RC4 streams [5], in order to strengthen RC4.

Sequence Counter (SC): The sequence counter stores a single natural number (initially zero) and increases by one each interval. The sender and each receiver have their own sequence counter (SC) in their memories.

For the sender, the SC number is increased by one for each new fixed-length data packet. For an encrypted fixed-length data packet, the data segment and MAC checksum are encrypted, but the SC number is not.

For each receiver, by checking the SC number of a new incoming fixed-length data packet, operations are performed as follows: if the incoming SC number is more than one integer value greater than the stored SC number, then the receiver updates its SC number to match the new value and decrypts the data segment by its CRS directly; otherwise, the receiver needs to compute the correct RC4 state by applying PRGA or IPRGA to its CRS a number of loops, and then to decrypt. The receiver has to update its SC number to match the new value which it received.

Figure 2: The (encrypted) fixed-length data packet

Fixed-length Data Packet: The proposed protocol requires that the length of the data packets which transmit between sender and each receiver is fixed. A fixed-length data packet includes a SC value, a data segment, and a MAC value. Note that if there are not enough data in the data segment of the last fixed-length data packet, the sender fills in a terminating symbol in the last packet, and adds random numbers to pad the empty place at the end of the data segment of the last fixed-length data packet.

3.2   Secure Data Transmission

This section presents the secure data transmission in the proposed protocol. During the handshaking phase (e.g. before the initial transmission, or the base key change), the sender and each receiver share the new RC4 base key, the MAC key, and the offset value O. Section 3.5 depicts MAC generation.

Note that, whereas WEP and WPA 1.0 reinitialize the RC4 state for every packet and generate the cipher stream from the initialized RC4 state, the proposed protocol does not reinitialize RC4 states frequently. Instead, the proposed protocol maintains the same RC4 base key for a duration known to each communicating node. This requires the initialization of the RC4 state (running KSA) to be carried out only when the base key changes. Furthermore, it is not necessary to update the offset value O frequently.

After acquiring the information during the handshaking phase, both sides apply KSA to generate an initial permutation of the RC4 state S_0, and then apply offset (O) loops of PRGA to S_0 to generate a new RC4 state as the CRS for key stream generation. At the same time, both sides set their sequence counters to zero (SC=0).

Figure 3 depicts the three-step operations which are executed by the sender. After these three steps, the encrypted fixed-length data packets are produced from the input plaintext message. The detailed descriptions of each step are presented below:

(1) The sender produces unencrypted fixed-length data packets. The sender divides the input plaintext message into contiguous fixed-length data segments and assigns the Sequence Counters (SC) to each of them. If there are not enough data in the data segment of the last fixed-length data packet, terminating symbol and random number (TR) as padding are added at the end of the data segment of the last fixed-length data packet for data encryption.

(2) The sender calculates the MAC checksum by inputting SC, data segments and the MAC key, and then fills the MAC checksum into the MAC segment. If there are not enough data in the data segment of the last fixed-length data packet, 0 are added at the empty place of the last fixed-length data packet for MAC generation. Section 3.5 depicts the MAC generation.

(3) The sender produces the encrypted fixed-length data packets. For each byte of the data which needs to be encrypted (here, it is the data segment and MAC segment), the sender XORs each byte with the correct stream key which is produced by the PRGA at the correct RC4 state. Following these operations for the bytes of the data which need to be encrypted, the encryption of the fixed-length data packet is completed and the packet is ready for transmission. The sender then increases its SC by one and updates its CRS.

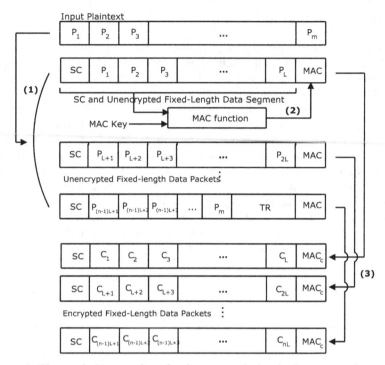

Figure 3: The sender's operations for data transmission in the proposed protocol

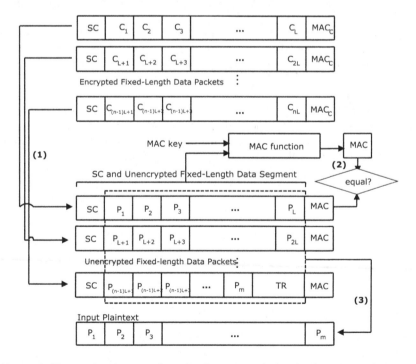

Figure 4: The receiver's operations for data transmission in the proposed protocol

Figure 4 depicts the three-step operations which are executed by the receiver. After the three steps, the input plaintext message is recovered from the corresponding fixed-length data packets. The details of each step are presented below:

(1) The receiver decrypts the encrypted fixed-length data packets. By reading the SC value of the coming fixed-length data packet, the receiver can find the right RC4 state based on the forward and backward property of RC4 states. Each receiver then calculates the correct key stream to decrypt the data which has been encrypted (here, it is the data segment and MAC segment) through XORing the corresponding key stream with the encrypted bytes.

(2) The receiver verifies the MAC checksum. The receiver verifies this fixed-length data packet by re-computing the MAC checksum. If the recomputed MAC checksum is the same as it was received, the MAC checksum is verified. Note that, 0 are added at the empty place of the last fixed-length data packet for MAC generation if there are not enough data in the data segment of the last fixed-length data packet.

(3) The receiver recovers the corresponding input plaintext message. After all the fixed-length data packets, which correspond to an input plaintext message, have been decrypted and their MAC checksum are verified, the receiver can restore them into the corresponding input plaintext message. The receiver then sets its SC as the SC value received, and updates its CRS.

3.3 Delayed or Lost Packets

In many network-based applications, it cannot be guaranteed that packets are received in the right order or there is no lost packet. Therefore, the application of a protocol which is based on the strong synchronization requirement is limited. Due to the backward and forward property of RC4 states, the proposed protocol allows that packets be received in an arbitrary order, or resending a lost packet.

3.4 Data Authentication

In our research, we assume that all communication nodes are trusted parties. Therefore, data authentication can be achieved through a purely symmetric mechanism, in which the sender and receiver share a secret key to compute a message authentication code (MAC) for all communicated data. When a message with a correct MAC arrives, the receiver knows that it was sent by the sender. If the receivers do not have strong trust assumptions, it needs an asymmetric mechanism. The asymmetric mechanism is a whole topic unto itself and is not covered in this paper.

No matter the receiver is trusted party or not, an MAC generation scheme is required for data authentication. MAC (message authentication code) is an authentication tag which is stored in the communication packet to verify the data authentication and data integrity during data transmission. Most popular method to compute MAC is to employ a secure hash function. Hash function-based MACs (often called HMACs) use a key or keys in conjunction with a hash function to produce a checksum that is appended to the message. There are a number of well known secure hash functions including SHA (SHA-512), MD5, Whirlpool and Ripemd-160. Some of them have been widely used and become key components of security protocols. However none of them is suitable for resource-constraint applications. For example, SHA-512 requires each block (1024 bits) to be processed by 80 rounds. Each round needs a large number of 64-bit applications [4]. In this paper, we present an RC4-based hash function in section 3.5 for MAC generation.

3.5. RC4-based Hash Function

In this section, we present an RC4-based hash function which can be used to generate MAC. Compared with the hash functions which are mentioned in section 3.4, the proposed hash function significantly reduces the computation overhead and achieves the requirements of a secure hash functions. Its security is based on the randomness of RC4 states. In the proposed RC4-based hash function, we use the notations of KSA* and PRGA*, each of which is a part of KSA or PRGA procedures that are listed below:

KSA*
for i = 0 to 255;
j = (j + S[i] + K[i]) mod 256;
swap (S[i], S[j]);

PRGA*
i=(i + 1) mod 256;
j=(j + S[i]) mod 256;
swap (S[i], S[j]);

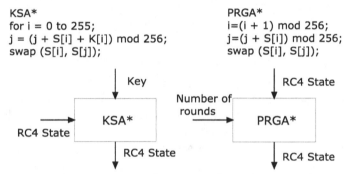

Figure 5: The operation codes of KSA* and PRGA* and their graphic representations

The input of KSA* is an RC4 state (e.g. an RC4 state which is generated by PRGA*) and a 64-byte key (K[0], K[1], ---, K[63]); the output is a new RC4 state. The input of PRGA* is an RC4 state and an integer to indicate how many loops the PRGA* applies; the output is another new RC4 state. Essentially KSA* is the same as KSA but it does not initiate the RC4 state at the beginning, and PRGA* is the same as PRGA but it does not generate the key stream.

The input of the proposed hash function is a message with a maximum length of less than 2^{64} bits and the output is a 258-byte message digest. The process of the proposed hash function can be described by the following two steps:

Step 1: Divide input message into 512-bit blocks.

The input message is divided into 512-bit blocks. The message is padded so that its length is congruent to 448 modulo 512 (length $\equiv 448$ mod 512). The padding bit is 0 and the number of padding bits is in the range of 0 to 512. Figure 6 depicts padding the input message. The output of this step is N pieces of 512-bit blocks notated by m_1, m_2, ---, m_N.

Note that when the proposed hash function is not used in the proposed protocol and the receiver does not have the length information of the input message, a segment for length information is required.

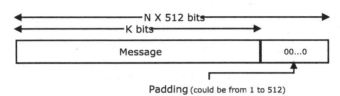

Figure 6: Padding the message

Step 2: Process message blocks.

The first 512-bit message block m_1 is processed by KSA and PRGA* in sequence. The input of the KSA is m_1 as the secure key and the output is an RC4 state named State$_{m1}$. The input of the PRGA* is State$_{m1}$ (as the initial state for PRGA*) and the output is an RC4 state named State$_1$ which is generated by applying (m_1 mod 2^5) loops of PRGA*. The output RC4 state (State$_x$) will be the initial RC4 state of KSA* to process next message block. The second 512-bit message block is processed by KSA* and PRGA* in sequence. The input of KSA* are the State$_2$ as the initial RC4 state and m_2 as the secure key and the output is State$_{m2}$. The same, State$_{m2}$ will be used as the input for the corresponding (m_2 mod 2^5) loops of PRGA* operation to generate State$_2$. The rest of blocks do the same process as the second block with corresponding input. State$_N$ is the 258-byte output of this RC4-based hash function. The process of this step is depicted in Figure 7 below.

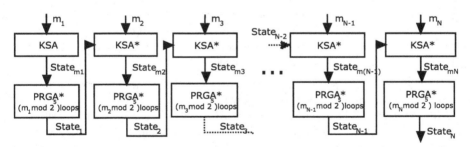

Figure 7: The procedure of the proposed RC4-based hash function

4   Analysis and Implementation

The proposed protocol is based on the RC4 algorithm with improvements and modifications. The improvements and modifications include using offset and the forward and backward property of RC4 states.

The forward and backward property of RC4 states is an intrinsic property of the RC4 algorithm. Any previous RC4 state can be recovered from the current or later RC4 state by simple addition and subtraction operations, as well as by swap operations. Since the proposed protocol simply uses this property and makes no changes to the original RC4, the proposed protocol does not reduce the security strength of RC4 by using the forward and backward property.

Offset is used in the proposed protocol to discard the first few bytes of the key stream, and data encryption or decryption begins at some later position. The purpose of using offset is to avoid the weaknesses of the initial RC4 states [3]. While the inclusion of offset is not a fundamental change to RC4, the security performance is improved. Thus, the proposed protocol does not reduce the security strength of RC4 by using offset.

For proposed RC4-based hash function, we have implemented a simulation to analysis the randomness of the RC4 state generation for KSA* and PRGA*. Simulation results show that the generation of the new RC4 state for KSA* and PRGA* maintains the same randomness as the KSA and PRGA.

Based on the analysis above, we believe that the proposed protocol uses RC4 in a way which makes effective use of RC4's strengths and reduces its weaknesses. The proposed protocol has achieved four main security properties required for designing a general secure communication scheme as below. Furthermore, the proposed protocol achieves semantic security and defends against lots of popular attack effectively, such as replay attacks and modified packet attacks.

- Data Confidentiality: The proposed protocol achieves data confidentiality through encrypting the data through the RC4 stream cipher.
- Data Authentication: The proposed protocol achieves data authentication through the MAC check.
- Data Integrity: The proposed protocol achieves data integrity through data authentication, which is a stronger way to check data integrity.
- Data Freshness: The proposed protocol achieves data freshness through the monotonically increasing value of the sequence counter for each data packet.

In the respect of the performance, the proposed protocol runs simply and efficiently because of the benefits of RC4 and some new features.

Using forward and backward property of RC4 states is one of the new features. By using it, the proposed protocol does not require key initialization frequently. This translates directly to some simplification and likely power saving, as well as improved performance over other original RC4 based protocols. For example, if the length of the data segment of the packet is 256 bytes, since KSA is unnecessary to be applied for every new packet, the encryption/decryption process of the proposed security protocol requires about half time than the time required by WEP.

Furthermore, all operations in the proposed protocol are simple swap, XOR, and addition and subtraction operations. Due to the simple implementation complexity, the proposed protocol is less expensive in terms of power and CPU resources. Considering the likely hardware support for RC4, it should be easy to selectively generate RC4 states from a base key or a pre-generated RC4 state. Hence, the proposed protocol should be compatible with existing hardware.

The proposed RC4-based security protocol has been implemented. It consists of two major parts. One is on sender side, and another is on receiver side. Our tests show that the proposed protocol works well to deal with the delayed or lost packet. Compared with the ordinary RC4 based security protocol, the proposed one are almost twice faster than the original one. Our tests also show that the proposed protocol is secure and has the ability to against a number of possible attacks including acknowledge attack, replay attack, and modified packet attack. We have applied the proposed protocol to multicasting environment where the numbers are dynamically changed. The experiment works correctly and smoothly. In the next step, we are going to implement the proposed protocol on a wireless sensor network which consists of a base station and a number of wireless sensor nodes. The results will be reported when the experiment is completed.

5   Conclusion

Wireless sensor networks and some of other resource-constrained applications are being deployed today and become more and more important. Security is a critical factor of these applications so a simple, efficient, and robust security protocol for secure data transmission on those devices is in large and urgent demand. This paper focuses on the designs of a lightweight security protocol for communications on wireless sensor networks which is based on the simple structure of the RC4 stream cipher and its forward and backward property.

Due to its simplicity, efficiency and high security, the proposed protocol is comparable with well-known security protocols. It is very simple to implement and efficient in both hardware and software, and is compatible with different levels of security. It can be widely used in many network applications. Furthermore, the proposed RC4-based hash function is a relatively independent part and can be served as one of the key components for other security applications.

References

[1] J. P. Kaps, G. Gaubatz, and B. Sunar. Cryptography on a Speck of Dust. IEEE Computer

Magazine, Feb. 2007, pp38-44.

[2] A. Perrig, R. Szewczyk, V. Wen, D. Culler, and J. D. Tygar. SPINS: Security Protocols for Sensor Networks. Wireless Networks, 2002, pp521-534.

[3] I. Mantin. Analysis of the Stream Cipher RC4. Master's thesis, The Weizmann Institute of Science, 2001.

[4] W. Stallings. Cryptography and Network Security, Principles and Practice. Prentice Hall, 2003.

[5] S. Mitchell and K. Srinivasan. State Based Key Hop Protocol: A Lightweight Security Protocol for Wireless Networks. In proceedings of the 1st ACM international workshop on Performance Evaluation of Wireless Ad Hoc, Sensor, and Ubiquitous Networks, 2004, pp112-118.

[6] A. Stubblefield, J. Ioannidis, and A.D. Rubin. Using the Fluhrer, Mantin, and Shamir Attack to Break WEP. AT&T Labs Technical Report, Aug. 2001

[7] S. Fluhrer, I. Mantin, and A. Shamir. Weakness in the Key Scheduling Algorithm of RC4. Proceedings, Workshop in Selected Areas of Cryptography, 2001.

Appendix

Theorem 1: For any r, $n \geq r \geq 0$, we have:

$IPRGA^r (S_n, i_n, j_n) = (S_{n-r}, i_{n-r}, j_{n-r})$

Proof. The proof is conducted by induction on integer r

Base Case:

It is true when r=0: $IPRGA^0 (S_n, i_n, j_n) = (S_n, i_n, j_n)$

Inductive Step:

Assume that for any $r \leq k$ we have: $IPRGA^k (S_n, i_n, j_n) = (S_{n-k}, i_{n-k}, j_{n-k})$

Consider case of $r = k+1$

$IPRGA^{k+1} (S_n, i_n, j_n) = IPRGA (IPRGA^k (S_n, i_n, j_n)) = IPRGA (S_{n-k}, i_{n-k}, j_{n-k})$

Let $IPRGA (S_{n-k}, i_{n-k}, j_{n-k}) = (S^*, i^*, j^*)$

According to PRGA and our notation, we have

$PRGA (S_{n-k-1}, i_{n-k-1}, j_{n-k-1}) = (S_{n-k}, i_{n-k}, j_{n-k})$ where $i_{n-k} = i_{n-k-1} + 1$ and $j_{n-k} = j_{n-k-1} + S_{n-k-1}[i_{n-k-1} + 1]$

$S_{n-k}[m] = S_{n-k-1}[m]$ where $m \neq i_{n-k-1}$ and $m \neq j_{n-k-1} + S_{n-k-1}[i_{n-k-1} + 1]$

$S_{n-k}[i_{n-k-1} + 1] = S_{n-k-1}[j_{n-k-1} + S_{n-k-1}[i_{n-k-1} + 1]]$

$S_{n-k}[j_{n-k-1} + S_{n-k-1}[i_{n-k-1} + 1]] = S_{n-k-1}[i_{n-k-1} + 1]$

According to IPRGA $(S_{n-k}, i_{n-k}, j_{n-k}) = (S^*, i^*, j^*)$,

we have $S^*[m] = S_{n-k}[m] = S_{n-k-1}[m]$ where $m \neq i_{n-k-1}$ and $m \neq j_{n-k-1} + S_{n-k-1}[i_{n-k-1} + 1]$

and $S^*[i_{n-k-1} + 1] = S_{n-k}[j_{n-k-1} + S_{n-k-1}[i_{n-k-1} + 1]] = S_{n-k-1}[i_{n-k-1} + 1]$

$S^*[j_{n-k-1} + S_{n-k-1}[i_{n-k-1} + 1]] = S_{n-k}[i_{n-k-1} + 1] = S_{n-k-1}[j_{n-k-1} + S_{n-k-1}[i_{n-k-1} + 1]]$

Thus $S^* = S_{n-k-1}$, $j^* = j_{n-k} - S^*[i_{n-k}] = j_{n-k-1} + S_{n-k-1}[i_{n-k-1} + 1] - S_{n-k-1}[i_{n-k-1} + 1] = j_{n-k-1}$, $i^* = i_{n-k} - 1 = i_{n-k-1}$

Therefore, we have $(S^*, i^*, j^*) = (S_{n-k-1}, i_{n-k-1}, j_{n-k-1})$

That is, for any $n \geq r \geq 0$, we have: $IPRGA^r (S_n, i_n, j_n) = (S_{n-r}, i_{n-r}, j_{n-r})$ □

HERO: Hierarchical kEy management pRotocol for heterOgeneous wireless sensor networks

Boushra Maala and Yacine Challal and Abdelmadjid Bouabdallah

Universite de Technologie de Compiegne, UMR CNRS 6599 Heudiasyc. France
(balkubei, ychallal, bouabdal)@hds.utc.fr

Abstract. Early researches focused on the security of homogenous sensor networks. However, recent works have demonstrated that the presence of heterogeneous sensor nodes gives better performance than homogenous ones in terms of energy consumptions, storage overhead, and network connectivity. In this paper, we propose a hierarchical key management scheme named HERO which is based on random key pre-distribution. HERO aims to construct a secure tree instead of complete connected graph as in existing schemes. Thanks to this realistic assumption, our key management scheme reduces considerably the number of pre-loaded keys assigned to each node while maintaining high security level at the same time. The preliminary simulation results using TOSSIM demonstrates that our scheme outperforms existing ones with respect to storage overhead.

1 Introduction

Recent advances in Wireless Sensor Networks (WSN) technology make them ideal candidates to various applications [5, 7] such as environment monitoring, military surveillance, forest fire monitoring, agriculture monitoring, etc. Security in WSN is a critical challenge especially for uncontrolled and hostile environments. Indeed, security vulnerabilities in WSN stem from their characteristics [8] that make them attractive: the wireless nature of radio channels exposes transmitted data in WSN to eavesdropping and manipulation, the resources limitations of motes prevent using heavy cryptographic techniques, and the deployment of sensors, generally, in unattended areas exposes them to intentional and accidental physical destruction. However, recent researches show that the performance and the lifetime of WSN can be improved by using heterogeneous sensor nodes instead of homogenous ones without significantly increasing the cost [9]. In Heterogeneous Sensor Networks (HSN) there is a few powerful and expensive nodes, and a large number of inexpensive and simple sensor nodes. A few key management schemes [11, 3, 4] have been proposed for (HSN). All these schemes try to exploit the heterogeneity features, in order to reduce the number of stored keys per node, and to guarantee a good level of security at the same time.

In this paper, we present an effective key management scheme based on probabilistic key pre-distribution; HERO(Hierarchical kEy management pRotocol for heterOgeneous wireless sensor networks). HERO provides a key management solution to secure a HSN relying on the construction of a secure tree. Moreover, it reduces the number of loaded keys because it needs only to store the keys which are necessary to build the secure tree, not to secure all possible links in the network as in all existing schemes[2, 11,

Please use the following format when citing this chapter:

Maala, B., Challal, Y. and Bouabdallah, A., 2008, in IFIP International Federation for Information Processing, Volume 264; Wireless Sensor and Actor Networks II; Ali Miri; (Boston: Springer), pp. 125–136.

3, 4]. The rest of the paper is organized as follows. In section II, we report on the related works presented in the literature. In section III, we detail our scheme. We evaluate the performances of HERO in section IV. In section V, we present some simulation results. Finally, we conclude the paper in section VI.

2   Related Works

Most of previous key management solutions have been proposed for homogeneous WSN, while a few works have been recently proposed for heterogeneous ones [11, 3]. The majority of these solutions are based on probabilistic key-distribution proposed by Eschenauer and Gilgor in [2]. This scheme is based on probabilistic key sharing among the sensor nodes. Prior to sensor network deployment, a large pool of P keys and their identifiers are generated. Each sensor node is pre-loaded with a key ring of k keys and their associated key identifiers which are selected randomly from the key pool without replacement. After sensor network deployment, each two neighbor sensor nodes compare their list of key identifiers in order to discover their shared key which will be used to secure the link between them. Due to the random distribution of keys to each sensor node, a shared key may be not available between two nodes. In this case, it is necessary to find a secure path composed of secure links using the shared keys of intermediary nodes. The probability of secure path existence using this scheme depends on the number of pre-loaded keys in each sensor. Unfortunately, to reach high probabilities this scheme may induce high storage overhead not supported by many types of sensor nodes. New key management schemes have been proposed for (HSN). These schemes are based on the basic scheme of probabilistic key pre-distribution [2]. Du et al. [11] proposed the Asymmetric Pre-distribution (AP) key management scheme. Its main idea is to pre-load a relatively large number of keys in each one of a small number of powerful nodes (H-sensor), while only a small number of keys are stored in each one of nodes (L-sensor) which have very limited space of storage and capacity of communication. Due to the usage of these two types of nodes, two different types of key rings have been used to achieve a high probability that each two nodes share at least one shared key. Indeed, AP scheme is more scalable than the basic scheme and it reduces the number of pre-loaded keys compared to the basic scheme, but it is still sometimes unsuitable for some types of sensors due to memory constraints. Also, in order to reduce the storage requirements and to maintain the same security strength of the basic scheme and AP scheme [2, 11], Hussain et al. [3] proposed a key generation process to reduce the key storage requirement. Brown et al. in [4], proposed a public-key-based scheme. In order to establish a secure communication topology in the network, each H-sensor needs to store (4+DH) keys where DH corresponds to the degree of this H-sensor. This means that each H-sensor stores enough keys to communicate with each node within its cluster, and the number 4 refers to four types of keys which are its public key, its private key, its group key, and the public key of the base station. For L-sensors, each node u must store (3+2Du) where Du corresponds to the group keys for all its neighbors, and the number 3 refers to three types of keys: its public key, its private key, and the public key of the base station. Although, this scheme may not need too large storage compared to other schemes, it is based on public key cryptography which is too heavy for tiny sensors. Recently, Traynor et al. [12] presented a protocol named LIGER. This protocol is based on the presence of a key distribution center(KDC), which stores the mapping of pre-loaded keys to nodes. This protocol consists of two components: the first is a protocol for a network in an infrastructure-less (i.e. without KDC). This component is defined as a protocol to instantiate probabilistic keying, and is referred to as LION. While, the second component is relied to the availability of KDC (i.e. network with KDC). It exploits the knowledge of the pre-loaded keys to perform the probabilistic authentication with a high degree of confidence. In addition, session keys are established with enforcement of the least privilege. This component is referred to as TIGER

All the schemes presented above assume the requirement to find a shared key between any two nodes in the network. We believe that this assumption is too strong. In fact, the primary goal in most WSN applications is to collect sensed information from the observed area, and not to have a complete connected graph. This can be achieved through a many-to-one communication scheme. Thus, we propose in our scheme HERO to elevate this assumption while achieving this primary goal. In HERO we replace the construction of a completely connected secure graph with a secure tree. As we can see in the performance evaluation sections, this allows to considerably reduce the required number of preloaded keys.

3   HERO(Hierarchical kEy management pRotocol for heterOgeneous wireless sensor networks) description

HERO is a new key management scheme that allows to create a secure tree rooted at the sink. Limiting the key sharing requirement to the child-parent relationship allows to reduce the key storage overhead to few keys per mote, as we can see in the performance evaluation section.

In HERO, we adopt a heterogeneous sensor network (HSN) consisting of a powerful base station (the Sink), and two types of sensor nodes: a small number of powerful sensor nodes with a large radio range (r), which are referred as Super nodes (Sn), and a large number of simple nodes with a small radio range (μ), which are referred as Normal nodes (Nn). Since sensor type influences the tree construction, each mote embeds to its sent messages an indicator (ind) that specifies the mote category (Sn or Nn). In the following paragraphs, we describe our scheme in details.

3.1   HERO Details

HERO scheme includes essentially two phases: Key pre-distribution and child-parent shared key discovery. In addition, we discuss two important operations in WSN: establishing keys for newly deployed sensors and the revocation of compromised keys.

Key Pre-distribution A large key pool (P) of P keys and the corresponding key Ids are generated off-line. Then, each Sn node is pre-loaded with K keys and their corresponding key Ids. These keys are randomly selected from the large key pool without replacement. We refer to this set of keys as KP (Key Pool), such that $KP_A \cap KP_B \neq \phi$ where A and B are two Sn nodes. Then each Nn node is pre-loaded with m ($m \ll K$) keys and their corresponding key Ids. These keys are randomly selected from P without replacement. We refer to this set of keys as PKP (Parcel Key Pool), such that $PKP_i \cap PKP_j \neq \phi$, and $PKP_i \cap KP_X \neq \phi$ where i and j are two Nn nodes and X is a Sn node.

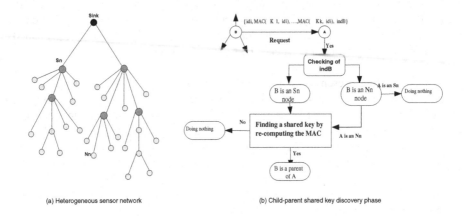

(a) Heterogeneous sensor network   (b) Child-parent shared key discovery phase

Fig. 1. Construction of a secure tree

Child-parent Shared Key Discovery The goal of HERO is to construct a secure tree (see figure 1). Therefore, HERO aims to construct a secure path between each node and the Sink, contrary to other schemes [11, 3] that aim to build a secure complete graph. After pre-loading each node with its key ring as described above, the Sink broadcasts a tree construction request. Since the Sink stores the key pool P, it is sure that all sensor nodes have at least one shared key with the Sink. When a sensor node (Sn or Nn) receives this request from the Sink, it considers the Sink as its parent, and selects one key of its key ring as its shared key with the Sink. These nodes are now attached to the tree rooted at the sink. Each newly attached node forwards the " tree construction request " downstream as follows:

Each attached node i (Sn or Nn) authenticates the request through computing one *Message Authentication Code* (MAC), over its identifier, using each key of its key ring. For instance, the "tree construction request" forwarded by node i will be:
$\{id_i, MAC(K1, id_i), MAC(K2, id_i), \ldots, MAC(Kk, id_i), ind_i\}$, where k is the size of node's key ring and ind_i refers to the type of node i. Then, node i broadcasts the request. When a node receives this message, it checks the type of the node. Thus, there are two different cases:

1. Receiving node is an Sn node:
   Suppose that A is the receiving node and B is the sending node. Then, A checks the type of node B. So:
   – If B is an Sn node. Then A checks the request message in order to discover if it has a shared key with B or not. This is done by verifying the corresponding

MAC. Here, if A finds a shared key with B, it notes node B as its parent and notes their shared key in its table of routing. Otherwise, A does nothing and it waits receiving a request from another node.
- If B is an Nn node. Then A waits wishing receiving a request from an Sn node.
2. Receiving node is an Nn node: If A finds a shared key with B, it notes B as its parent in its routing table and their shared key. Otherwise, it waits receiving another request.

Establishing keys for newly deployed sensors The deployment of new nodes in WSN may be required for several reasons such as: Solving the problem of the miss of coverage in the tested area, enhancing the accuracy of the sensed phenomena, and prolonging the lifetime of the deployed network.

To attach a new node A to the secure tree, the new node must search for a parent, already attached to the secure tree, with whom it shares a key. Thus, A is firstly pre-loaded with its key ring of k keys which are selected randomly from the key pool. Then, it broadcasts a request message as follows:

$\{Id_A, MAC(K1, id_A), MAC(K2, id_A), \ldots, MAC(Kk, id_A), ind_A\}$.

When a deployed sensor node B, which is already an attached node, receives this message, it broadcasts the following message:

$\{Id_B, MAC(K1, id_B), MAC(K2, id_B), \ldots, MAC(Kk, id_B), ind_B\}$.

When A receives this message, it checks if it has a shared key with B. If A and B have a shared key, A notes B as its parent. Otherwise A waits receiving a message from another node. Furthermore, due to one of the reasons which are cited above, a deployed node C may be still unattached to the secure tree. So, upon receiving the request of the new node A, C attempts to profit from this new node to attach to the secure tree. Thus, it checks if there is a shared key with A, if a shared key is found, C considers A as its parent.

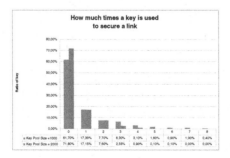

Fig. 2. Frequency of key usage with different key pool sizes

Key Revocation Whenever a sensor node is compromised, it is necessary to revoke all its keys. First, the sink generates a signature key that it sends securely to each sensor except the compromised one. One method to distribute this signature key securely to the sensors [2], is to encrypt it with a list ok keys selected from the key rings of not compromised sensors, so that all sensors could decrypt the signature key except the compromised one. Then the sink broadcasts a message containing the identifiers of the compromised keys authenticated with the signature key. Upon receiving this revocation message, each sensor node verifiers the signature using the signature key. Then, it removes the corresponding keys from its key ring. However, due to the revocation of the compromised keys, some links may disappear. So, in order to reconfigure these links; the affected nodes restart a new child-parent shared key discovery phase.

Figure 2 shows that, out of P=1000 keys, only 60% of the keys are used to secure links, only 17.30% are used to secure one link, 7.70% are used to secure two links, and only 6.30% are used to secure three links. This means that compromising one key leads to the compromise of another link with probability 0.17, of two other links with probability 0.07 and so on. Moreover, the compromise of one Nn node of key ring size m=5 leads to reconfigure only 4 links in average. So, a very small number of nodes are affected due to the compromise an Nn. Therefore, only a small number of messages are required to reconfigure the affected links.

4   Evaluation of HERO

In order to construct a secure tree with a minimum key storage overhead, we must find the minimum size of an Nn's key ring such that the probability of that Nn attaches to the secure tree be high. An Nn node attaches to the secure tree as a child of an Sn node except when this Nn node has not any shared key with at least one Sn node within its neighborhood. In this case, as we have explained in previous section, the Nn node attaches to the secure tree as a child of another Nn node within its neighborhood with which it has a common key. Therefore, the probability that an Nn node attaches to the secure tree (P_a) is the probability that an Nn node shares at least one key with at least one Sn within its neighborhood (P_{Sn}) or that it does not share any key with any Sn node within its neighborhood $(1 - P_{Sn})$ and it shares at least one key with at least one Nn within its neighborhood (P_{Nn}). Thus, we can write this as:

$$P_a = P_{Sn} + (1 - P_{Sn}) \times P_{Nn} \qquad (1)$$

Where, P_{Sn} is the probability that an Nn node shares at least one key with at least one Sn of its neighborhood, then P_{Sn} = 1- the probability that no key in common between an Nn node and any one of Sn node within its neighborhood. Then,

$$P_{Sn} = 1 - \left[\frac{\binom{P}{K+m}\binom{K+m}{m}}{\binom{P}{m}\binom{P}{K}} \right]^{N_{Sn}} \qquad (2)$$

where N_{Sn} is the number of Sn nodes within a Nn node's neighborhood.

Similarly, since P_{Nn} is the probability that an Nn node shares at least one key with at least one Nn within its neighborhood, then, P_{Nn} =1- the probability that no key in common between an Nn node and any one of Nn node within its neighborhood. Then,

$$P_{Nn} = 1 - \left[\frac{\left(\frac{P}{2 \times q}\right)\left(\frac{2 \times q}{q}\right)}{\left(\frac{P}{m}\right)^2} \right]^{N_{Nn}} \tag{3}$$

where N_{Nn} is the number of Nn nodes within a Nn node's neighborhood. Thus, from 1, 2 and 3, and by using *Stirling approximation* $n! \simeq \sqrt{2\pi n}.(\frac{n}{e})^n$, we obtain:

$$P_a \simeq H^{N_{Sn}} + V^{N_{Sn}} \times U^{N_{Nn}} \tag{4}$$

where:

$$H = 1 - \sqrt{1 + \frac{m \times K}{P^2 - P \times (m+K)}} \times \left(1 + \frac{m \times K}{P^2 - P \times (m+K)}\right)^P \times \left(1 - \frac{K}{P-m}\right)^m \times \left(1 - \frac{m}{P-K}\right)^K$$

$$V = \sqrt{1 + \frac{m \times K}{P^2 - P \times (m+K)}} \times \left(1 + \frac{m.K}{P^2 - P \times (m+K)}\right)^P \times \left(1 - \frac{K}{P-m}\right)^m \times \left(1 - \frac{m}{P-K}\right)^K$$

and

$$U = 1 - \frac{(1 - \frac{m}{P})^{2(P-m+1/2)}}{(1 - \frac{2m}{P})^{(P-2m+1/2)}}$$

Let us assume that $N = n_1 + n_2$ (n_1 and n_2 being the numbers of Sn and Nn nodes respectively, where $n_1 \ll n_2$) are uniformly distributed within a square area having a surface $Z = L \times L\ m^2$. Thus, the density of deployed Sn and Nn nodes are $D_{Sn} = \frac{n_1}{Z}$ and $D_{Nn} = \frac{n_2}{Z}$ respectively. Then, the possible number of Sn node within a zone which is covered by an Nn node is $N_{Sn} = D_{Sn} \times Z_{Nn} = \frac{n_1 \times \pi \times \mu^2}{Z}$, where μ is the radius of Nn node. Also, the possible number of Nn nodes within a zone which is covered by an Nn node is $N_{Nn} = D_{Nn} \times Z_{Nn} = \frac{n_2 \times \pi \times \mu^2}{Z}$.

Hereafter, we present an example of a network consisting of 10 IRIS motes as Sn nodes with radio range r=350m, and 1000 MicaZ motes as Nn nodes with radio range μ=90m. These sensor nodes are deployed uniformly within a square-grid of $500 \times 500 m^2$ with density $D = \frac{N}{Z} = 40.4 \times 10^{-4}$ node/m^2. Then, the possible number of Sn and Nn nodes within an Nn node's neighborhood are: N_{Sn}=1 node and N_{Nn}= 102 nodes respectively. By using the equation 4, where P=3000, m=5 and K=100, we obtain Pa=0.6397. Thus, we notice that an Nn node needs only to be pre-loaded with a very small number of keys to be attached to the secure tree with a relatively high probability.

Figure 3 shows the probability of attachment Pa as function of m and K i.e. Pa=f(m,K) where m={1,2,3,4,5} and K={100,150,200,250,300}.

Figure 3.a illustrates Pa=f(m,K) for P=1000, while figure 3.b illustrates Pa=f(m,k) for P=3000. However, by comparing these two figures, we can clearly notice that Pa decreases when P increases when m and K are fixe. For instance, with K=100 and m=3 we obtain Pa= 0.7097 for P=1000 and 0.335 for P=3000. Moreover, Pa increases when K increases when m and P are fixe. For example, with P=1000 and m=1, there

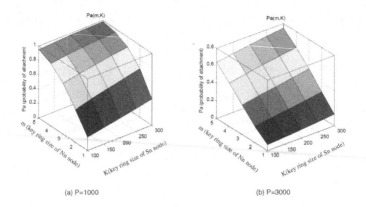

(a) P=1000   (b) P=3000

Fig. 3. Pa=f(m,K) where K={100,150,200,250,300}, m={1,2,3,4,5}, $N_{Sn} = 1$, N_{Nn}=102

is an increment of 24% in Pa when K increases from 100 to 150. We notice also that Pa increases when m increases when K and P are fixe. For instance, for P=1000 and K=100, there is an increment of 54% in Pa when m increases from 2 to 3. Thus, we can conclude that with a very small size of m, HERO can give a high probability of attachment, and hence a high probability of constructing a secure tree. HERO can also be considered as an efficient protocol since it provides a high security level (compared to existing protocols) with a very small key storage overhead.

5 Simulation and analysis

We have implemented a prototype of HERO using TinyOS [6] environment. TinyOS is an event-driven operating system commonly used on WSN motes. The programs are written in nesC which is a C-like language [10, 14]. Basically, we have defined a new interface KeyManager, and developed a module implementing this interface KeyManagerM. It contains all commands and events which are necessary to manage the keys in a network. We have also developed the component TreebuilderM that uses KeyMangerM to build the secure tree.

In what relates to the network topology, we considered a square-grid of $500 \times 500m^2$. Super nodes have a 350m radio range (like IRIS motes), and Normal nodes have a 90m radio range (like MicaZ motes) [13]. The positions of the nodes are distributed uniformly over the square grid. We assumed also that 1% of the nodes are Super nodes. We carried out intensive simulations of our prototype using TOSSIM [1] which emulates the execution of the developed code over a personal computer. We were mainly interested in evaluating the secure coverage ratio with respect to the network size, and the key ring size at Super Nodes and Normal Nodes for different key pool sizes. We mean by the secure coverage ratio the number of motes that succeed to establish a secure path to the Sink, over the total number of motes in the network.

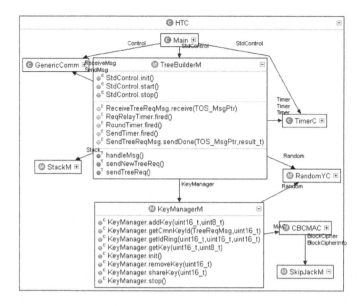

Fig. 4. The components and the interfaces of TinyOS used to implement HERO

5.1 Secure coverage ratio vs. key ring sizes

In this simulation scenario, we considered a 1000 motes network. Mote 0 being the Sink, and 10 motes are considered as Super nodes. We considered different key ring sizes for two cases: a pool of 1000 keys and a pool of 2000 keys. Figure 5 illustrates the evolution of the secure coverage of the network for a 1000 and 2000 key pool respectively. We notice that in both cases the secure coverage of the network increases when increasing the key ring sizes. What is interesting to notice is the fact that with small key rings we can reach very high secure coverage of the network. Namely for a 1000 key pool, we can reach more than 90% of secure coverage with only 7 keys at Nn nodes and 50 keys at Sn. Depending on the resources available at Super nodes, we can reach the same coverage ratio while further decreasing Nn ring size: we can use 6 keys in Nn ring while using 100 keys in Sn ring, or 5 keys in Nn ring while using 150 keys in Sn ring, or only 4 keys in Nn ring while using 200 keys in Sn ring. We also notice that the smaller is the key pool, the greater is the coverage ratio. However, the rekeying overhead is smaller when the key pool is greater. Indeed, with a large key pool only few nodes share the same keys of a compromised node (cf. 3.1). Table 5.1 compares HERO to some solutions of the literature with respect to the probability of key sharing. We notice that HERO induces the smallest key ring storage overhead while achieving the highest key sharing probability.

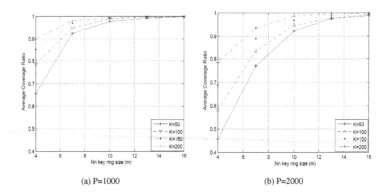

Fig. 5. Evolution of secure coverage ratio with respect to key rings sizes

Table 1. Comparison of some key pre-distribution solutions with respect to key sharing probability for Key pool size P=1000

WSN type	Protocol	Parcel key pool size(K)	Key ring size(m)	Probability of sharing a key
Homogeneous	Eschenauer and Gligor's basic scheme	100	100	0.5
Heterogeneous	Asymmetric key pre-distribution scheme (AP)	500	20	0.5
	Brown et al. public-key based scheme	104	43	0.5
Heterogeneous	**HERO**	**100**	**3**	**0.7**

5.2 Secure coverage ratio vs. network size

In this simulation scenario, we considered a key pool size of 1000 keys. Each Super Node is initiated with a ring of 100 keys, and each Normal Node is initiated with a ring of 5 keys. The key rings are randomly chosen from the 1000 key pool. Then we varied the network size from 100 nodes to 1000 nodes. For each point, we run our simulation 10 times and we calculated the average secure coverage ratio of the network. Figure 6 illustrates the evolution of secure coverage ratio when we increase the network size. We notice that the secure coverage ratio depends on the density of the network: the greater is the network density, the higher is the secure coverage ratio. Moreover, for network size in $[100 - 300]$ nodes the secure coverage ratio increases with almost 100%. With only 550 motes, HERO allows to reach more than 90% of secure coverage of the network. This means that with only 1 mote per $450m^2$ we reach more than 90% of secure coverage using HERO.

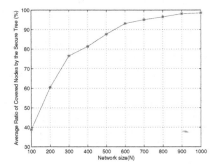

Fig. 6. Evolution of secure coverage ratio with respect to network size

6   Conclusions

In this paper, we have studied random key pre-distribution for wireless sensor networks. This category of key management schemes aims to secure links between sensors of a WSN. Many studies and applications demonstrate that communications in WSN follow mostly many-to-one paradigm. Also, the new works confirm that, the usage of heterogeneous nodes gives better performance than homogenous ones in terms of energy consumptions, storage overhead, and network connectivity. Therefore, we exploited the heterogeneity features, in this paper, to propose a new key management scheme that secures this communication paradigm through constructing a secure tree rooted at the Sink. Our solution (HERO) is designed for heterogeneous sensor networks and relied on random key pre-distribution. HERO secures child-parent links through pairwise shared keys. Preliminary simulation results using TinyOS show that our solution reduces the key storage overhead compared to existing schemes.

References

1. Levis, P., Lee, N., Welsh, M., Culler, D. : TOSSIM: Accurate and Scalable Simulation of Entire TinyOS Application. In: ACM SenSys. Los Angeles(2003)
2. Eschenauer, L., Gligor, V.D. : A key management scheme for distributed sensor networks. In: Proceedings of (CCS '02.), pp. 41-47. Washington(2002)
3. Hussain, S., Kausar, F., Masood, A.: An efficient key distribution scheme for heterogeneous sensor networks. In: the international conference on wireless communications and mobile computing, pp. 388-392.Hawaii(2007)
4. Brown, J., Du, X., Nygrad, K. : An Efficient Public-Key-based Heterogeneous Sensor Network Key Distribution Scheme. In: IEEE GLOBECOM. Washington(2007)
5. Jason Lester H. : System Architecture for Wireless Sensor Network. In:PhD Dissertation, University of California, Berkeley (2003)
6. Hill, J., Szewczyk, R., Woo, A., Hollar, S., Culler, D., Pister, K. : System Architecture Directions for Networked Sensors. In: Proceeding of ACM ASPLOS IX. (2000)
7. Xu, N. : A Survey of Sensor Network Applications. In: Survey Paper for CS694a. Computer Science Department, University of Southern California (2002)

8. Camtepe, S. A., Yener, B. : Key Distribution Mechanisms for Wireless Sensor Networks: a Survey. In: TechReport TR-05-07. System Design Laboratory (2005)

9. Yarvis, M., Kushalnagar, N., Singh, H., Rangarajan, A., Liu, Y., Singh, S. : Exploiting Heterogeneity in Sensor Networks. In: IEEE INFOCOM'05. Miami(2005)

10. Gay, D., Levis, P., Behren, R., Welsh, M., Brewer, E., Culler, D. : The nesC language: A Holistic Approach to Networked Embedded System.In: Precedings of (PLDI).California (2003)

11. Du, X., Xiao, Y., Guizani, M.,Chen,H-H.: An effective key management scheme for heterogenous sensor networks. Ad Hoc Networks, Elsevier. 5(1), 24–34 ,Jan (2007)

12. Traynor, P., Kumar, K., Choi, H. : Effcient Hybrid Security Mechanisms for Heterogenous Sensor Networks. IEEE Trans.(MTC). 6(6), 663–677,Jun (2007)

13. Technology. Crossbow : http://www.xbow.com

14. UC. Berkeley : http://nescc.sourceforge.net

Designing incentive packet relaying strategies for wireless ad hoc networks with game theory

Lu Yan and Stephen Hailes

University College London, Dept. of Computer Science
Gower Street, London WC1E 6BT, UK
{l.yan | s.hailes}@cs.ucl.ac.uk

Abstract. In wireless ad hoc networks, nodes are both routers and terminals, and they have to cooperate to communicate. Cooperation at the network layer means routing (finding a path for a packet), and forwarding (relaying packets for others). However, because wireless nodes are usually constrained by limited power and computational resources, a selfish node may be unwilling to spend its resources in forwarding packets that are not of its direct interest, even though it expects other nodes to forward its packets to the destination. In this paper, we propose a game-theoretic model to facilitate the study of the non-cooperative behaviors in wireless ad hoc networks and analyze incentive schemes to motivate cooperation among wireless ad hoc network nodes to achieve a mutually beneficial networking result.

1 Introduction

In a wireless ad hoc network, intermediate nodes on a communication path are expected to forward packets of other nodes so that nodes can communicate beyond their wireless transmission range. Typical examples of the wireless ad hoc networks include military communication or emergency response scenarios, where all participating nodes belong to the same authority, share a common goal, and are therefore motivated to act cooperatively to provide network services [1]. However, since wireless ad hoc networks are increasingly deployed in daily life scenarios such as inter-vehicle communication [2] and Internet access for remote areas [3], the participating nodes most unlikely belong to a single authority and no longer share a common goal any more. Cooperation in such networks cannot be assumed since it has been commonly accepted that network nodes in the above scenarios have the freedom to make decisions in their own best interests.

Since wireless nodes are usually constrained by limited power and computational resources, a selfish node may be unwilling to spend its resources in forwarding packets which are not of its direct interest, even though it expects other nodes to forward its packets to the destination. This is particularly true in typical wireless sensor network nodes [4], where battery-power is a scarce resource and forwarding is an energy-consuming network activity that will shorten a node's lifetime. Even in an ideal case in which nodes have no power and computational constraints, forwarding packets consumes a portion of bandwidth and processing time available to the

Please use the following format when citing this chapter:

Yan, L. and Hailes, S., 2008, in IFIP International Federation for Information Processing, Volume 264; Wireless Sensor and Actor Networks II; Ali Miri; (Boston: Springer), pp. 137–148.

forwarding nodes. Thus, the forwarding cost is not zero; a reasonable move for a node which does not belong to a central authority is to drop the packets belonging to any other node but itself, i.e., to act selfishly. In the worst case, assuming every node is selfish, this behavior will have a collective effect to bring down the communication mechanism of the whole ad hoc network, and result to a *tragedy-of-the-commons* [5].

In this paper, we propose a game-theoretic model to facilitate the study of the selfish behaviors in wireless ad hoc networks and analyze incentive schemes to motivate cooperation among wireless ad hoc network nodes to achieve a mutually beneficial networking result. The rest of this paper is organized as follows. Section II reviews the literature. In Section III we study two classical strategies under the game theoretic framework. We present a novel incentive strategy for wireless nodes in Section IV and then conclude the paper in Section V.

2 Related work

Non-cooperative issues in ad hoc networks have drawn considerable attention over the past few years, and it has been shown that the presence of selfish nodes degrades the overall performance of a non-cooperative ad hoc network [6].

Ideally, a right amount of incentives should be provided to motivate cooperation among wireless ad hoc network nodes to achieve a mutually beneficial networking result. Briefly, participating nodes should be rewarded for cooperative behaviors and be punished for non-cooperative behaviors. The incentive schemes for packet forwarding in the literature basically fall into two categories, namely, trust-based schemes and price-based schemes.

Trust-based schemes use trustworthiness and reputation information in routing to enforce the cooperation among nodes. In [6], participating nodes are designed to perform monitoring to overhear the packet retransmission and avoid transmission to misbehaving nodes. CONFIDANT [7] distributes trustworthy information among those participating nodes, and every node is supposed to keep a reputation list of all its previous interacted nodes. A subjective logic trust model is proposed in [8] for the mapping between the evidence space and the trust space, and further applied to the trusted protocol for wireless ad hoc networks. This design is similar in [9], but with more technical details and data results. More recent approaches such as [10] and [11] evaluate trust level of a node by aggregating feedback information from its neighbors to reduce the communication overhead. STRUDEL [12] is a distributed framework that tackles the problem of free-riders in Coalition Peering Domains by using a Bayesian trust model to dynamically select and isolate malicious peers.

Price-based schemes introduce services charges to the packet transmission process and usually use micro-payment to compensate the resource consumption incurred from the transmission. A virtual currency NUGLET [13] is introduced by L. Buttyan and J.-P. Hubaux as an economic incentive proposal. Each intermediate node buys a packet for some nuglets, and sells it to the next one for more nuglets. Therefore a node increases its nuglet amount during packet forwarding. SPRITE [14] applied the same idea, but with a credit based system and a central clearing banking service, to alleviate the need to carry the real currency. A priced priority forwarding scheme was studied in [15], where as pricing mechanism allows the nodes to arbitrarily set the cost of priority forwarding of a packet. APE [16] is a virtual economy scheme to encourage the intermediate nodes to reveal the cost of packet forwarding, and thus to choose a cost-efficiency route. A general discussion about micro-payment schemes is available [17], but not specific to the wireless ad hoc networks.

Since a node's behaviors in wireless ad hoc networks mirror the real world decision making process, game theory models and designs have been applied in the wireless ad hoc networks literature. In [18], V. Srinivasan et al. proposed a game playing strategy to achieve Nash equilibrium which converges to the rational and Pareto optimal. The fairness issue was studied in [19] for bandwidth allocation in ad hoc networks and a game theory model for intrusion detection in wireless ad hoc networks was discussed in [20]. A survey on applying game theory to wireless ad hoc networks is available [21]. Some of the basic analysis of forwarding in single-stage and repeated-game scenarios can be founded on [28, 29]. However, the existing works are either not explicit or too general. In this paper, we propose a formal action model of wireless ad hoc network nodes based on game theory. Since all nodes have to periodically choose an action (e.g. forwarding or dropping), this is a repeated game for all the participants. Because of the complexity of the problem, we restrict our analysis to a static network scenario in this paper.

3 Game theoretic model

Let us first consider the simplest forwarding scenario which consists of two nodes i and j, where i wants j to forward a packet to other nodes. The denotations used in the following discussions are summarized in Table 1.

Table 1. Denotation list

r	Reward
s	Resource
b	Benefit
t	Time
d	Discount rate
f	Forward probability
$-i$	Node's opponent
h	Tolerance threshold

3.1 The single-stage game

We assume that j has freedom to decide whether to forward or drop the packet. We also assume there is a reward mechanism such that for each successful forwarding, i will get its reward of value r for enforcing j to forward packets; and meanwhile, j will consume its resource of value s to complete the forwarding.

Further, we define the interaction model between nodes as bi-directional, i.e., while j tranmits i's packet, i shall transmit j's packet simultaneously. Since bi-directional wireless communication is never physcially simultaneous, this setting has implication on the physical layer in which the two transmissions in one interaction should be independent, i.e., nodes should make decisions on forwarding or dropping at the same time.

We can see the above setting is a single stage prisoner's dilemma [22], where the motivation for a node to participate in this game playing is to maximize its benefit and the action it can take is to decide whether to forward or drop. For this game, the benefits accrued by each node for every strategy profile are tabulated in Table 2.

Table 2. Benefit vs. strategy

	Forward	Drop
Forward	r-s, r-s	-s, r
Drop	r, -s	0, 0

In this typical single stage prisoner's dilemma, the best strategy for a participating node is to drop the packet, since this action will maximize its benefit. We can further prove that mutual dropping will lead to Nash equilibrium: since the benefit function of each node is monotonically increasing with dropping actions and the maximized benefit of a node is r, dropping is a dominating strategy in this case. In other words, no node can gain more profit by cooperation with others. Putting into the wireless ad hoc networking context, it implies that selfish actions are actually encouraged but this will lead to zero throughput of the whole network.

3.2 The repeated game

If the packet forwarding game is played only once, there is no way we can achieve cooperation among these participating nodes; but in a real world setting, all nodes have to periodically choose an action (e.g. forwarding or dropping), and it becomes a repeated game for all the participating nodes. Similarly to the single stage game above, it achieves a non-optimum equilibrium when played repeatedly; however, sub-optimum equilibriums are achievable, provided that players do not know a priori how many repetitions of the game there will be.

Since a repeated game is path-dependent, a node has to take consideration of the present action's impact on its future consequence [23]. We extend our previous model to reflect the impact of past event on future actions: assume that time t is discrete and divided in frames like t_1, t_2 .. t_n. A node i will make a decision at each stage t_n of the game, but its benefit gained at each stage will be discounted by a rate at d to reflect a past action's impact factor over time [24].

A participating node in this game takes actions by adjusting its attitude towards forwarding. Explicitly, we denote a node's probability to forward a packet as f, where in this case $f=0$ means dropping all the requested packets and $f=1$ mean forwarding all the requested packets. Let i denote the playing node's id and $-i$ denote its opponent, the benefit of a node for playing this game at a time frame t_n is $b_i^{t_n} = f_{-i}^{t_n} r - f_i^{t_n} s$. The final benefit of a node finishing the game is

$B_i = \sum_{n=0}^{\infty} d^n (f_{-i}^{t_n} r - f_i^{t_n} s)$, where d is the discount rate. We study two classical

strategies [23] and their sub-optimum equilibrium conditions.

3.2.1 Tit For Tat

TFT (Tit For Tat) is a highly effective strategy in game theory for the iterated prisoner's dilemma [23]. A player using this strategy will initially cooperate, and then respond in kind to an opponent's previous action. If the opponent previously was cooperative, the player is cooperative. If not, the player is not. We model the TFT strategy as follows:

$$f_i^{t_n} = \begin{cases} 1, & n = 0 \\ f_{-i}^{t_{n-1}}, & n > 0 \end{cases} \tag{1}$$

TFT provides incentives to cooperate since one player's present move will have impact on its future consequence. Let's study the node dynamics in this strategy.

Assume all participating nodes are adopting TFT at the bootstrapping stage. It is obvious that this cooperation will continue as the forwarding probability will remain 1 and thus all packets will be forwarded. If at some point a node i unilaterally changes its forwarding probability to $f_i^{t_0} = p$ (e.g. due to a physical transmission failure), its opponent will copy its behavior and set its forwarding probability to $f_{-i}^{t_1} = p$; but the node i itself follows its opponent's previous behavior, thus its own forwarding probability will be set back to $f_i^{t_1} = 1$. The above process will continue repeatedly and in the end we will see an alternate changing sequence as follows:

$$f_i^{t_n} = \{p, \ 1, p, \ 1, \ \cdots\} \tag{2}$$
$$f_{-i}^{t_n} = \{1, p, \ 1, p, \ \cdots\}$$

We can calculate the final benefit of node i for playing this game:

$$\begin{aligned} B_i &= \sum_{n=0}^{\infty} d^n (f_{-i}^{t_n} r - f_i^{t_n} s) \\ &= [(r - ps) + d(pr - s)] * (1 + d^2 + d^4 + \cdots) \\ &= \frac{(r - ps) + d(pr - s)}{1 - d^2} \end{aligned} \tag{3}$$

A node will adopt this strategy if and only if it will gain profit from it, i.e. $B_i \geq 0$. We solve Equation 3 and derive the boundary condition as:

$$\frac{r}{s} \geq \frac{d+p}{1+dp} \tag{4}$$

3.2.2 Grim Trigger

Grim Trigger (GT) is a trigger strategy in game theory for a repeated game [23]. Initially, a player using GT will cooperate, but as soon as the opponent defects (thus satisfying the trigger condition), the player using GT will defect for the remainder of the iterated game.

Since a single defect by the opponent triggers defection forever, GT is the most strictly unforgiving of strategies in an iterated game. The implication in the network design is that once a node is misbehaved, it will be isolated permanently. We model GT as follows:

$$f_i^{t_n} = \begin{cases} 1,\ n = 0 \\ 1,\ \forall m < n: f_{-i}^{t_m} \geq h \text{ and } n > 0 \\ 0,\ \text{all else} \end{cases} \tag{5}$$

where h is introduced as the tolerance threshold and the trigger condition is $f < h$.

We will then discuss the sub-optimum equilibrium condition for GT. Suppose all nodes are adopting GT at the bootstrapping stage, and the cooperation continues until some point when a node i unilaterally changes its forwarding probability to $f_i^{t_0} = p$. Depending on the degree of deviation, its opponent will respond with GT: (1) a slight degree of deviation, i.e. $p \geq h$, will be tolerated; and it will still forwards all packets (2) a bigger deviation, i.e. $p < h$ will trigger the punishment, and it will from now on drop all the packets forever. If we consider the worst case $p < h$, the forwarding probability sequence will be

$$f_i^{t_n} = \{p,\ 1, 0, 0,\ \cdots\} \tag{6}$$
$$f_{-i}^{t_n} = \{1,\ 0, 0, 0,\ \cdots\}$$

Let us calculate the final benefit of node i for playing this game:

$$B_i = \sum_{n=0}^{\infty} d^n (f_{-i}^{t_n} r - f_i^{t_n} s) \tag{7}$$
$$= r - ps - ds$$

Using the condition $B_i \geq 0$, we get the boundary condition for a node to apply this strategy:

$$\frac{r}{s} \geq d + p \tag{8}$$

4   Proposed incentive strategy

Any breach of cooperation in TFT or GT results in either all packets being dropped or some fraction of packets being dropped. In other words, fully cooperative is never the Nash equilibrium point with TFT or GT. Meanwhile, the boundary conditions from Section III imply that cooperation can be achieved in a network of selfish nodes, given network parameters (e.g., earning-cost ratio) carefully designed. Besides, in a real world communication process, network fluctuations and measurement errors may trigger unjust punishments, and those punishments may have irrevocable future effects in some strategies; in designing a practical strategy for real world networking, recovery mechanism shall be taken into serious considerations.

From the above analysis, we can see that a strategy without recovery mechanism is not stable in practice. We aim to design a new strategy that reciprocates both cooperation and defection, but shall still be able to rebuild cooperation after a node's unintentional misbehaviors.

It is our hypothesis that an optimal strategy should bear at least the following merits:

--It is good (it starts by cooperating)
--It is retaliating (it returns the opponent's defection)
--It is generous (it forgets the past if the defecting opponent cooperates again)
--It is not memoryless (it utilizes the history information).

Our goal is to propose a strategy that is more adaptive to fluctuations and able to re-achieve full cooperation after a node's misbehaviors. For the case of wireless ad hoc networks we design the Gradual strategy based on the idea of adaptiveness [25]: a player uses cooperation on the initial move and then continues to do so as long as its opponent cooperates. Then after the first defection of the opponent, it defects one time and cooperates two times; after the second defection of the opponent, it defects two times and cooperates two times, ... after the Nth defection it reacts with N consecutive defections and then calms down its opponent with two cooperations:

$f_i^{t_0} = 1$; N = 0; n = 1; // start with forward

```
while(n++) { // time-series

    if( f_{-i}^{t_{n-1}} < h ) { // opponent drop

        N++; M = n + N;

            while(n < M + 1) // respond with N drops
```

$$\{ \, f_i^{t_n} = 0 \, ; \quad \texttt{n++;} \, \}$$

$$f_i^{t_n} = 1 \, ; \quad \texttt{n++;} \quad f_i^{t_n} = 1 \, ; \quad \texttt{// two forwards}$$

$$\}$$

$$\texttt{else} \quad f_i^{t_n} = 1 \, ; \quad \texttt{// otherwise forward}$$

$$\}$$

Using the same assumptions and settings in analyzing the previous strategies, we can derive a forwarding sequence after a node's misbehavior as follows:

$$f_i^{t_n} = \{p, \, 1, \, 0, \, 1, \, 1, \, \cdots\} \tag{9}$$
$$f_{-i}^{t_n} = \{1, \, 0, \, 1, \, 1, \, 1, \, \cdots\}$$

We shall admit this is the simplest case with Gradual, and Equation 9 may become very complex when a node/opponent misbehaves again. Though we can still use the previous mathematics analysis method to find out the sub-optimum equilibrium for Gradual, the boundary condition itself shall take a complex form whose implication is no longer intuitive. Thus it is our hypothesis that simulation is a better way to understand the dynamics in this strategy.

In a real world networking environment, nodes may take divergent strategies. For instance, (1) a group of nodes may employ TFT, and the other may use GT; (2) a node may take TFT in the bootstrapping stage, but it may switch to GT in a later stage. In other words, we need to study the mixed-strategy situations. However, such situations are complex in nature and simulation may be a better way to understand the cooperation behavior over time, which is generally believed to be *evolutionary* [26].

Complex strategies in the iterated prisoner's dilemma have been studied by Axelrod [26] and a computer simulated tournament was proposed to evaluate different strategies when they compete against each other [27]. Typically, game theorists use a round robin tournament to study the behavior of strategies. Each strategy meets all other strategies in this tournament, and its score (i.e., benefit) is recorded for each confrontation. Its final score is the un-discounted sum of all scores, and the strategy's goodness is measured by its final score completing the tournament.

However, the above simulation design only provides the final result and does not reveal any dynamic aspects of strategy behaviors throughout the simulation process. We adopt the idea of ecological evolution and revise the round robin tournament as follows: we model the mixed strategy scenario as an ecological evolution process; at the beginning there is a fixed population of the same quantity of each strategy; a round robin tournament is made and then the population of bad strategies is decreased while good strategies obtain new elements. The simulation is repeated until the population has been stabilized. In this simulation scenario, a good strategy is then the strategy which stays alive in the population for the longest possible time, and in the biggest possible proportion [30].

Suppose that, initially, the total population is composed of M strategies S_i where $i = 1, \ldots, M$. At the generation n, each strategy is represented by a certain number of players as $W^n(S_i)$; Let $V(S_i \mid S_j)$ denote the score (i.e. benefit) when strategy S_i meets strategy S_j, which can be derived, for instance, from Table 2. The computation of the score $g^n(S_i)$ of a player using a selected strategy S_i at the generation n is:

$$g^n(S_i) = \sum_{j=1}^{M} (W^n(S_j) V(S_i \mid S_j)) - V(S_i \mid S_i) \qquad (10)$$

The size of each sub-population $W^{n+1}(S_i)$ at the generation $n+1$ is:

$$W^{n+1}(S_i) = \left\lfloor \frac{W^n(S_i) g^n(S_i) \sum_{j=1}^{M} W^n(S_j)}{\sum_{j=1}^{M} (W^n(S_j) g^n(S_j))} \right\rfloor \qquad (11)$$

where all divisions shall be rounded to the nearest lower integer, due to the physical meaning of $W^n(S_i)$.

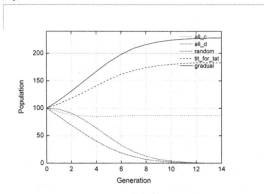

Fig. 1. Simulation result

The simulation consists of 500 random nodes with an even distribution of 5 strategies: `all_c` (always cooperate), `all_d` (always defect), `random`, `tit_for_tat`, `gradual`, i.e., initially each strategy is represented by 100 players. We run the simulation algorithm described in section 4.4.1, and plot the population changes over time in Fig. 1. It is possible to observe from Figure 1, after 12 generations the `all_d` and `random` strategies disappear; `gradual` has the biggest portion of population, followed by `tit_for_tat`. This implies that Gradual strategy outperforms TFT and proves to be the winning strategy in a mixed strategy environment.

We have then repeated the simulation with more strategy sets and all results show that Gradual is relatively strong, compared to other strategies; and the most significant point is that Gradual has never, or not often, bad scores. Therefore it is our conjecture that Gradual is a good strategy for wireless nodes.

The reason why Gradual outperforms others can be explained as follows: different from all previous memory-less strategies, Gradual keeps a long-term memory about nodes' behaviors and adaptively punishes misbehaviors with history information. Meanwhile it is positive towards rebuilding cooperation with two consecutive cooperations after the punishment.

Actually we can find a number of real life metaphors of this strategy in our daily life, e.g. credit agency vs. debtors, government vs. taxpayers. A node's standing in this strategy can therefore be interpreted in two ways:

- Offensive: a node wants to force its opponent to cooperate, and thus clearly shows that it will be more and more aggressive so that the best choice for its opponent is to cooperate.
- Defensive: a node does not want to be exploited, and thus gets less and less for cooperation. However as it needs services from others, it retries sometimes to reinstall cooperation.

5   Concluding remarks

In this paper, we have proposed a game theoretic model to investigate the conditions for cooperation in wireless ad hoc networks. Because of the complexity of the problem, we have restricted ourselves to a static network scenario. The main finding is that a cooperative state in a network of selfish nodes without a central authority is theoretically achievable, given network parameters (e.g. cost-earning ratio, discount rate, tolerance threshold, etc) carefully designed [31].

We have then studied different node strategies within this model, and derived boundary conditions of cooperation for those strategies. As one of the results, we have proposed a refined strategy which is adaptive to fluctuations and able to achieve a full cooperation after a node's unintentional misbehaviors. Due to the memory effect of this strategy, we conducted a simulation study instead of mathematical analysis. Another significant finding in this paper lies in the evaluation of mixed strategy scenarios, where different nodes adopt different cooperation strategies [32].

Throughout the paper we have taken the assumption that packet transmission is unicast. However, recently there has been a plethora of work on using multicast routing protocols in wireless ad hoc networks [33]; we also intend to investigate the packet dynamics in a multicast network and extend the model for it.

Acknowledgment

This work is partly funded by EPSRC under grant EP/D07696X/1.

Reference

1. L. Buttyan and J.-P. Hubaux, "Stimulating Cooperation in Self-organizing Mobile Ad Hoc Networks," in ACM/Kluwer Mobile Networks and Applications, vol. 8, no. 5, pp. 579-592, Oct. 2003.
2. T. Nadeem, S. Dashtinezhad, C. Liao, L. Iftode, "TrafficView: traffic data dissemination using car-to-car communication", in ACM SIGMOBILE Mob. Comput. Commun. Rev., Vol. 8, No. 3, pp. 6-19, July 2004.
3. C.E Perkins, E. M. Belding-Royer, and Y. Sun, "Internet connectivity for ad hoc mobile networks", in International Journal of Wireless Information Networks, April 2002.
4. S. Kumar, A. Arora, T.H. Lai, "On the lifetime analysis of always-on wireless sensor network applications", in Proc. IEEE International Conference on Mobile Ad hoc and Sensor Systems, 2005.
5. G. Hardin, "The Tragedy of the Commons," Science, Vol. 162, No. 3859, pp. 1243-1248, December 1968.
6. S. Marti, T. J. Giuli, K. Lai, and M. Baker, "Mitigating Routing Misbehavior in Mobile Ad Hoc Networks", in Proc. ACM/IEEE International Conference on Mobile Computing and Networking (Mobicom), Boston, August 2000.
7. S. Buchegger and J.-Y. Le Boudec, "Performance analysis of the CONFIDANT protocol," in Proc. International Symposium on Mobile Ad Hoc Networking & Computing (MOBIHOC 2002), Lausanne, Switzerland, June 2002.
8. X. Li, M.R. Lyu, J. Liu, "A trust model based routing protocol for secure ad hoc networks", in Proc. IEEE Aerospace Conference, 2004.
9. T. Ghosh, N. Pissinou, K. Makki, "Towards designing a trusted routing solution in mobile ad hoc networks", in Mobile Networks and Applications, Volume 10, Issue 6, December 2005.
10. Q. He, D. Wu and P. Khosla, "SORI: A Secure and Objective Reputation-based Incentive Scheme for Ad hoc Networks," in Proc. of IEEE Wireless Communications and Networking Conference (WCNC2004), Atlanta, GA, USA, March 2004.
11. M. T. Refaei, V. Srivastava, L. DaSilva, and M. Eltoweissy, "A Reputation-based Mechanism for Isolating Selfish Nodes in Ad Hoc Networks," in Proc. IEEE Second Annual International Conference on Mobile and Ubiquitous Systems: Networking and Services (MOBIQUITOUS 2005), San Diego, CA, July 2005.
12. D. Quercia, M. Lad, S. Hailes, L. Capra and S. Bhatti, "STRUDEL: Supporting Trust in the Dynamic Establishment of peering coalitions", in Proc. of ACM Symposium on Applied Computing SAC 2006, Dijon, France, April 2006.
13. L. Buttyan and J.-P. Hubaux, "Enforcing service availability in mobile ad-hoc wans", in Proc. IEEE/ACM Workshop on Mobile Ad Hoc Networking and Computing (MobiHOC), Boston, MA, 2000.
14. S. Zhong, J. Chen, and Y. R. Yang, "Sprite: A Simple, Cheat-Proof, Credit-Based System for Mobile Ad Hoc Networks," in Proc. of IEEE Infocom 2003, San Francisco, CA, USA, April 2003.
15. B. Raghavan and A. C. Snoeren, "Priority forwarding in ad hoc networks with self-interested parties", in Proc. Workshop on Economics of Peer-to-Peer Systems, Berkeley, CA, 2003.
16. L. Anderegg and S. Eidenbenz, "Ad hoc-VCG:A Truthful and Cost-Efficient Routing Protocol for Mobile Ad hoc Networks With Selfish Agents", in Proc. ACM/IEEE International Conference on Mobile Computing and Networking (Mobicom'03), San Diego, CA, 2003.
17. M. Peirce, "Multi-party Micropayments for Mobile Communications", PhD Thesis, Trinity College Dublin, Ireland, Oct. 2000.
18. V. Srinivasan, P. Nuggehalli, C. F. Chiasserini, and R. R. Rao, "Cooperation in wireless ad hoc networks," in Proc. IEEE INFOCOM 2003, San Francisco, CA, Mar./Apr. 2003.

19. Z. Fang and B. Bensaou, "Fair bandwidth sharing algorithms based on game theory frameworks for wireless ad-hoc networks", in Proc. IEEE INFOCOM 2004.

20. A. Patcha and J.-M. Park, "A Game Theoretic Formulation for Intrusion Detection in Mobile Ad Hoc Networks", in International Journal of Network Security, Vol.2, No.2, Mar. 2006.

21. V. Srivastava, J. Neel, A. B. MacKenzie, R. Menon, L. A. DaSilva, J. E. Hicks, J. H. Reed, and R. P. Gilles, "Using game theory to analyze wireless ad hoc networks," in IEEE Communications Surveys and Tutorials, vol. 7, pp. 46-56, 2005.

22. P. K. Dutta, "Strategies and Games: Theory and Practice," MIT Press, 1999.

23. G. J. Mailath and L. Samuelson, "Repeated Games and Reputations: Long-Run Relationships", Oxford University Press, 2006.

24. M. S. Morgan, D. F. HendryNet, "The Foundations of Econometric Analysis", Cambridge University Press, 1997.

25. B. Beaufils, J.-P. Delahaye and P. Mathieu, "Our Meeting With Gradual: A Good Strategy For The Iterated Prisoner's Dilemma", in Proc. the Fifth Int'l Workshop on the Synthesis and Simulation of Living Systems, MIT Press, 1997.

26. R. Axelrod and W. D. Hamilton, "The evolution of cooperation", Science, 211: 1390-1396, 1981.

27. R. Axelrod, "The evolution of cooperation", New York, 1984.

28. L. A. DaSilva and V. Srivastava, "Node Participation in Ad-hoc and Peer-to-peer Networks: A Game-theoretic Formulation", Workshop on Games and Emergent Behavior in Distributed Computing Environments, September 18, 2004, Birmingham, U.K.

29. F. Milan, J. J. Jarmillo, and R. Srikant, "Achieving Cooperation in Multihop Wireless Networks of Selfish Nodes", Workshop on Game Theory for Networks, October 14, 2006, Pisa, Italy.

30. M.A. Nowak, A. Sasaki, C. Taylor, D. Fudenberg, "Emergence of cooperation and evolutionary stability in finite populations", Nature, 428: 646-650, 2004.

31. L. Yan, S. Hailes, L. Capra, "Analysis of packet relaying models and incentive strategies in wireless ad hoc networks with game theory", in Proc. IEEE 22nd International Conference on Advanced Information Networking and Applications (AINA'08), Okinawa, Japan, 2008.

32. L. Yan, S. Hailes, "Cooperative Packet Relaying Model for Wireless Ad hoc Networks", in Proc. ACM International Workshop on Foundations of Wireless Ad Hoc and Sensor Networking and Computing (FOWANC'08), co-located with ACM MobiHoc'08, Hong Kong, China, 2008.

33. Z. Li, "Min-Cost Multicast of Selfish Information Flows", in Proc. of the 26th Annual IEEE Conference on Computer Communications (INFOCOM'07), Anchorage, Alaska, May 6-12, 2007.

Energy Efficient Key Management Protocols to Securely Confirm Intrusion Detection in Wireless Sensor Networks

Jibi Abraham and K S Ramanatha
Department of Computer Science & Engineering
M S Ramaiah Institute of Technology, MSRIT Post,
Bangalore - 560 054, Karnataka, India
Email: jibiabraham@msrit.edu, ksramanatha@msrit.edu

Abstract: Wireless Sensor Networks are prone to many security attacks. The most complex among them is the node compromise attack. Networks enhanced with services like aggregation and security require a different intrusion detection mechanism than the generally used solutions and there is a possibility of a compromised node producing false intrusion detection alarms. Therefore we need suitable mechanisms to detect intrusion and to securely confirm the same. We propose two major schemes: 1) a post-deployment key distribution based permanent key establishment scheme 2) two on-the-fly key establishment schemes. The simulation results and analysis show that on-the-fly schemes are better suited for intrusion confirmation in energy constrained sensor networks. A simple and practical intrusion defense scheme also is suggested.

Keywords: Sensor networks, intrusion detection, secure confirmation of intrusion, key management.

1   Introduction

Wireless sensor networks (WSN) are vulnerable to different types of attacks [9] and many of these attacks survive only in sensor networks because the deployment of the network might be in inaccessible terrain and providing costly tamper resistant hardware is not an expected feature. In a node compromise attack [1], an attacker seizes a node from the sensor network, connects this node to his laptop, extracts the stored data, puts new data/behavior and has control on that node. The node compromise attack can generate many other security attacks [9][10] and the effect of all these attacks finally leads to either unavailability of the network functionality or short network life.

In this paper we consider a network that is divided into many clusters. The cluster head in each cluster manages the cluster memberships, the distribution of the security key to members and aggregation of the collected data from the members as described in paper [7]. The data is encrypted at the cluster member by using a pair-wise key or cluster-wide key and decrypted at the cluster head. The cluster head performs aggregation on the collected data, encrypts the data using a different key belonging to the next hop or its parent cluster and the encrypted data is sent to the next node. Detection of attacks in these networks requires a specialized Intrusion Detection Scheme (IDS). Further, even after using the special IDS, a compromised node can

---

Please use the following format when citing this chapter:

Abraham, J. and Ramanatha, K.S., 2008, in IFIP International Federation for Information Processing, Volume 264; Wireless Sensor and Actor Networks II; Ali Miri; (Boston: Springer), pp. 149–160.

always send false intrusion detection alarms about legitimate nodes in the network. Hence proper key management protocols are required to confirm intrusion after detection.

In this paper we propose an intrusion detection scheme as well as a scheme to securely confirm the detection. This is done in Section 3. Section 2 brings out rather limited attempts in this direction in the literature. The intrusion confirmation scheme requires establishment of pair-wise keys between the intrusion detector and the nodes on the data aggregation tree path. In Section 4 we bring out the unsuitability of key pre-distribution method for pair-wise key establishment [3][4][5] in malice prone sensor networks. We have modified and enhanced the grid-based polynomial pre-distribution scheme of [3] to design a new scheme that we have called post-deployment permanent key establishment scheme suitable for secure confirmation of intrusion. We conduct detailed evaluation to bring out deficiencies of such schemes, in general. We therefore propose two on-the-fly schemes which make use of our original work in [7] for the establishment of pair-wise keys. A comparison of all the three schemes is made to bring out the superiority of on-the-fly schemes in general and on-the-fly scheme using the Base Station in particular. In Section 5, a suitable   intrusion defense scheme is suggested.

In this paper emphasis is placed on the key management protocols required to confirm intrusion and therefore on their implementation and simulation results. An analysis of the secured data thus collected needs to be made to determine whether the intrusion has really taken place or it is a case of a false alarm and for identifying the intruder in the case of intrusion. However such a confirmation analysis is not within the scope of this paper. Further, the intrusion detection as well as the defense mechanisms are only proposed. Their implementation and evaluation are left for future work.

2   Literature Survey

Even though a few watchdog mechanisms [1][2] are available in the literature, they are not suitable for the networks that support aggregation and security. A secure aggregation scheme is suggested in [6], where the data forging attack at a specific aggregator can be detected. Each node A has a pair-wise key K_{AS} shared with the Base Station S, stored as a pre-deployment knowledge. In the ith data transmission phase, a leaf node A computes a temporary key $K_{i-AS} = E_{KAS} (i)$, sends its data reading R_A, node ID, ID_A and Message Authentication Code, $MAC (R_A, K_{i-AS})$ to its parent. Here $E_{KAS} (i)$ denotes the encryption of i using the secret key K_{AS}. The parent node B calculates the aggregation $Aggr$ of its children nodes' readings and sends $Aggr$, ID_B and $MAC (Aggr, K_{i-BS})$ to its parent. When the final aggregation result reaches the Base Station, the Base Station verifies the final aggregate and during data validation phase it broadcasts the temporary keys (K_{i-AS}, K_{i-BS},...) to the network. Using the temporary keys received, an intermediate aggregator verifies the intermediate aggregation results. Even though the scheme achieves low data communication through aggregation, it is vulnerable because the intermediate aggregation can be easily tampered if a parent and a child node in their hierarchy are compromised. Also data validation step consumes lots of battery power by broadcasting the temporary keys and re-computing the MAC.

It is observed that schemes available in the literature for intrusion detection and its secure confirmation for sensor networks offering aggregation and security services are very limited and are inadequate. The newly proposed schemes of this paper detect intrusion posed by a single compromised node as well as intrusion from a set of cooperative nodes and avoid chances of any false intrusion detection alarms.

3   Detection and Confirmation of Misbehavior in Sensor Networks Offering Aggregation and Security Services

Some of the major misbehaviors that are resulted from a compromised node are explained first. In *Spoofing* attack, a node supplies falsified sensed data. In a *Masquerade* attack, a compromised node supplies false data as if the data gets originated at a non-compromised node. In *Falsified Aggregate* attack, a compromised cluster head computes the aggregate of the data supplied by its members and instead of relaying the actual aggregate results, it will transmit an altered aggregate to its parent. Many pair-wise key establishment protocol steps include message traversal between one or more intermediate nodes and there is a possibility of many of the intermediate compromised nodes breaching the protocol steps by playing **man-in-the-middle** attack.

We propose a new intrusion detection mechanism that avoids unnecessary buffering of bulk data received from many nodes. Usually multiple sensors trigger event detection in a sensor network and all the sensors may report the same detection data to the Base Station. Assume that every cluster is numbered and each data message includes the cluster number of the data origin, data source ID and the aggregated data value. When a data originated from a cluster reaches at a cluster head on the data traversal path at the first time, the cluster head CH_k at level k inserts a record including cluster number and the aggregate value into a Cache. Since there is not much possibility of wide deviation in values sensed by neighbor nodes for a specific event detection, when the data arrives from the same cluster at later times, if CH_k finds wide deviation in the stored aggregate value with the latest received aggregate, then it suspects the possibility of alteration attack at CH_{k+1}, the next lower level.

After perceiving deviation, the cluster head that has detected intrusion intimates the Base Station and as part of the intrusion defense mechanism, the Base Station may instruct all the other nodes in the network to reduce communication towards the suspected node. But here comes possibility of another attack like a compromised cluster head sending false intrusion detection alarm about a legitimate neighbor cluster head to the Base Station. As a result of the defense mechanism, all the data communications towards the legitimate cluster head may get diverted to the compromised cluster head and this compromised cluster head may start misbehaving on the data. Hence it is always better to confirm the intrusion by directly contacting all the nodes on the data traversal path of the deviated aggregate value. This leads to designing an *Intrusion-Confirm* protocol.

The *Intrusion-Confirm* protocol consists of a request message and a reply message. Let the data originator be the source node S_i. To verify and confirm the occurrence of an attack, CH_k prepares an *Intrusion-Confirm* request message demanding to provide the value of the actual data supplied by cluster of S_i. This request message is sent to all the nodes on the aggregation tree path starting from CH_{k+1} to S_i. After collecting *Intrusion-Confirm* reply messages from all nodes, CH_k analyzes and confirms whether CH_{k+1} is a compromised node or not.

There can be a possibility of a sudden actual shift in the sensed data. In this situation the immediate cluster head at first notices the change. If any other cluster head other than the immediate cluster head perceive the value shift then that may be a case of an intrusion. Even if two cluster heads make a collaborative attack, since the aggregation corresponds to the data from the source cluster, attack can be detected at the higher level. The possibility of a compromised cluster head/member masquerading a legitimate cluster head/member by supplying false data can also be detected. The scheme requires data buffer to store the *Intrusion-Confirm* reply messages and data messages from many neighbour nodes only after the detection to confirm the intrusion.

The *Intrusion-Confirm* request and reply messages are to be secured because otherwise the malicious node in between can alter the request and reply messages by playing man-in-the-middle attack. Therefore there is a requirement of pair-wise key establishment from CH_k to all the nodes in the aggregation tree path from CH_{k+1} to S_i which is the main concentration of this paper.

4   Pair-wise Keys Establishment for Intrusion Confirmation

Schemes for pair-wise key establishment can be categorized into centralized and decentralized. In centralized schemes, the Base Station acts as a Key Distribution Centre to establish pair-wise keys after verifying the authenticity of each node, assuming that Base Station is physically guarded and can never get compromised. Decentralized schemes based on key pre-distribution [3][4][5], claim to be the most suitable for sensor networks. The nodes in the network co-operate among each other and establish the pair-wise keys. But in a malice prone network, any sensor node may get compromised at any time or fresh malicious nodes may also get added to the network at any time. All the key pre-distribution schemes actually permit any node to establish keys with other nodes without verifying the authenticity, if they have the appropriate pre-distribution material.

The grid-based polynomial pre-distribution scheme presented in [3] claims that the scheme has high probability to establish pair-wise keys, is tolerant to key capture and has low communication overhead. So we decided to try the possibility of using this scheme to establish the required pair-wise keys to confirm intrusion. The scheme utilizes t-degree bi-variate polynomials $f(x, y) = \sum_{i,j=0}^{t} a_{ij} x_i y_j$ over a finite field F_q, where q is a large prime number and satisfies the property that $f(x, y) = f(y, x)$. The scheme uses a 2-dimensional grid of size $m \times m$, where $m = \lceil \sqrt[2]{N} \rceil$ and N is the number of nodes in the network. Each row i of the grid is associated with a row-polynomial $f_{i,r}(x, y)$ and each column i is associated with a column-polynomial $f_{i,c}(x, y)$. Each node has to be assigned a grid ID (i, j) that is a row and column intersection point in the grid so that the node stores the polynomial shares of ith row and j^{th} column. A pair-wise key is established either as a Direct-Key or as an Indirect-Key. Two nodes can establish a Direct-Key if both of them share the same row or column polynomial. Otherwise, nodes need to find a key path such that any two adjacent nodes in the path can establish a Direct-Key and by utilizing the nodes on the key path an Indirect-Key is established.

The proposed protocol to confirm intrusion requires usage of pair-wise keys to securely contact the nodes on the data traversal path. If the polynomial based pre-distribution scheme is used after intrusion detection, there is possibility of man-in-the-middle attack posed by one of the compromised nodes included in the key path if an Indirect-Key has to be established. So the required pair-wise keys cannot be established on-the-fly, but they are to be established permanently in the initial phase of the network formation assuming that it not possible to compromise any node at the initial phase of network formation. Hence we have proposed a post-deployment permanent key establishment scheme.

4.1 Post-deployment Permanent Key Establishment scheme

The grid-based scheme in [3] is modified and enhanced as follows to design a post-deployment permanent key establishment algorithm:

- The basic work assumes that each node in the network is in communication range with all other nodes in the network. If a node cannot have Direct-Key with a second node, that node itself can decide the key path required to establish Indirect-Key. But this is not the situation in real time deployment of sensor networks, where not all nodes are in communication range with each other and they use multi-hop communication protocol to communicate each other. Sensor nodes may not be having enough resources to store the details about all the other nodes in the network and the key path discovery process has to be very similar to a route discovery used to establish a route between two nodes that may introduce substantial communication overhead. Hence the basic scheme is enhanced with the Base Station storing the details about all the nodes existing in the network. Whenever any node requests for a key path to another node, the Base Station returns it if a successful path is available.
- The basic work provides a method to convert node address to Grid-ID, which is called as *row-wise* assignment. It assigns consecutive addresses to the Grid-ID starting from the first row and the first column, row-wise. i.e. in the order of $(0,0)$, $(0,1)$, . . .,$(0, m-1)$, $(1, 0)$, . . ., $(1, m-1)$ and so on up to $(m-1, m-1)$ where m is the dimension of the grid. But if the network is designed for a larger value of m but lesser number of nodes is used in the actual deployment, not all row polynomials get well utilized in the scheme. An improved Grid-ID assignment called *Diagonal-wise* assignment is proposed, in which consecutive addresses are assigned in a diagonal way so that all the polynomials are equally well distributed.
- As there are chances of physical capture of nodes from the deployment location to extract the polynomial shares stored at the node, it is preferred to distribute the polynomial share after network deployment, as part of node authentication to the Base Station. The Base Station generates $2m$ number of polynomials immediately after its initialization. As soon as a node self-organizes and joins the network, it will register to the Base Station. The Base Station stores the details about each registering node in a Register and the details include a list of registered nodes with which this currently registering node can establish a Direct-Key. The Base Station returns the row and column polynomials to that node, if the registration is successful.

When a node gets connected to the network and recognizes its parent node, it establishes pair-wise keys with its parent and all the grandparent nodes on the aggregation hierarchical tree using the enhanced Grid-based scheme. The key establishment steps are given in Table1.

Table 1. Permanent Key establishment algorithm

Step	Description
1	Register the node ID to the Base Station and obtain the polynomial shares
2	If Direct-Key is feasible to its parent • From the matched row or column polynomial, derive the key. • Send a Direct-Key confirmation message to the parent. • Parent verifies the Direct-Key and returns the node ID of its parent, if the parent is not the Base Station.
3	If Indirect-Key is feasible to its parent • Request the Base Station for key path to the parent. • When the key path is obtained, generate a random pair-wise key. • Using the nodes on the key path, the encrypted pair-wise key is intimated to the parent. • The parent verifies the pair-wise key and returns the node ID of its parent, if the parent is not the Base Station.
4	If Direct-Key is feasible to the next grand parent • From the matched row or column polynomial, derive the pair-wise key. • Send a Direct-Key confirmation message to the grand parent. • The grand parent verifies the Direct-Key and returns the node ID of its parent, if the parent is not the Base Station.
5	If Indirect-Key is feasible to the next grand parent • Request the Base Station for key path to the grand parent. • When the key path is obtained, generate a random pair-wise key. • Using the nodes on the key path, encrypted pair-wise key is intimated to the grand parent. • The grand parent verifies the key and returns the node ID of its parent, if the parent is not the Base Station.
6	Repeat the steps 4 and 5 until no more grandparents exist for this node.

4.1.1   Protocol Evaluation

The protocol is implemented for TinyOS [11] using NesC language. RC5 algorithm is used for encryption/decryption operations. The simulations are done in TOSSIM [8] for various sized networks and results are taken. While the protocol certainly achieves the main purpose of secure confirmation of intrusion detection, it is beset with several drawbacks. Therefore, a very detailed evaluation is done to bring out the deficiencies of the scheme in real time implementation. The following observations are made:

•   The maximum eligible payload size of any *TinyOS* message is 29 Bytes. The structure of the Indirect-Key establishment message used in the implementation, its field interpretations and memory requirement for each field are given in Table 2. It is clear that a maximum of 5 nodes can only be included in the key path to frame an eligible TinyOS message. During simulation of the proposed protocol, it is observed that the length of the key path formed was 2, 3 and 4 for a maximum of 250 nodes in the network. So if the Grid-based scheme is used for pair-wise key establishment with more than 250 nodes then there is a possibility of key establishment failure because it may not be possible to find a key path with a maximum of 5 nodes.

Table 2. Structure of Indirect Key Establishment Message

Message-field	Interpretation	Size (Byte)
Message-type	Key establishment with parent or grand parent	1
Key-path [5]	List of nodes on the key path	10
Key [16]	Encrypted pair-wise key, which has to be established	16
Path-length	Length of the key path	1
Current-node	Node in the key path where currently processing is done	1

• As the length of key path increases, the overhead towards computation and communication also increases. If the key path length is four, the Indirect-Key is established with the help of three intermediate nodes, requiring four times hop-to-hop transmissions, four times encryptions and four times decryptions. The energy expense towards pair-wise key establishment with the parent nodes and with the grand parent have been evaluated. The energy expense is directly proportional to the number of keys established and more energy is spent if the number of Indirect-Keys is more than the number of Direct-Keys. The evaluations are taken for both *Row-wise* and *Diagonal-wise* Grid-ID assignments. It is observed that when the networks size is less than 100, the number Direct-Keys are more in *Row-wise* assignment compared to *Diagonal-wise* assignment. But for larger size networks, *Diagonal-wise* scheme has more number of Direct-Keys. The percentage of Direct-Keys formed with parents the grand parents, compared to the total number of keys formed in the network is shown in Figure 1 and Figure 2 respectively. It is observed that for larger networks, the percentage of Direct-Keys is drastically reduced as the number of nodes in the network increases.

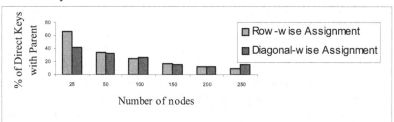

Fig. 1. Percentage of Direct-Keys formed with parents

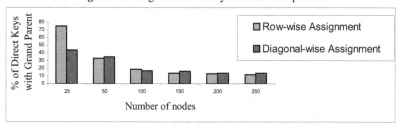

Fig. 2. Percentage of Direct-Keys formed with Grand parents

- The hierarchical trees formed during the simulations are with unequal degree for different sub trees. The maximum degree of these trees is between 6 and 8. As the number of nodes in the network increases, the number of levels in the trees is also increasing. The number of levels in the trees in the simulations with the *Diagonal-wise* assignment is given in Figure 3. The number of keys to be stored at the Base Station is same as total number of nodes in the network. But the number of keys to be stored in many nodes in the next few levels down the Base Station is also very near to number of keys stored at the Base Station. The Base Station can be a high-end node, but all the other nodes in the network are normal resource constrained sensor nodes. The number of keys being stored at the nodes in level 0, level 1 and level 2 in the hierarchical trees of various sizes are shown in Figure 4. As the number of keys needs to be stored at the nodes is high, the memory overhead of the protocol is very high.

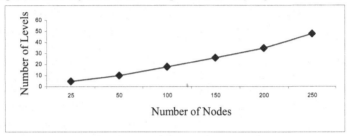

Fig. 3. Number of levels in the hierarchical trees with *Diagonal-wise* assignment

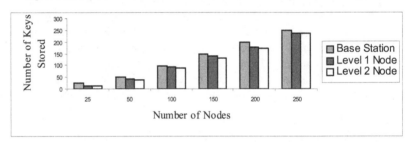

Fig. 4. Highest number of keys stored in various levels

- When a node requests for an Indirect-Key path, The Base Station finds out a key path from the details of list of registered nodes. Many times the Base Station could not find a successful key path because Grid-IDs are assigned either row-wise or diagonal-wise in increasing order of address of the nodes and node registration is done in a random order. So at a particular time, there is a possibility of unavailability of a common node to establish the key path, consequences to a failure in finding a key path and the requesting node is required to repeat the key path request until it could get one. The number of times the key path request has failed for the entire network is also evaluated and the details are given in Figure 5. As the number of nodes increases, the number of failures also increases.

The above protocol evaluation has brought out the deficiencies of permanent key establishment protocols, in general. But it has been observed that as the network size increases, the memory required for key storage increases and also the communication overhead is more due to massive Indirect-Key formation as it is clear from Figure 1 and Figure 2. Sensor networks are very much resource constrained and therefore the memory required for key storage has to be reduced. Further there is no likelihood that all the nodes in the network surely get compromised. Therefore most of the keys established through the permanent scheme do not get utilized as the network gets compromised sparingly in its lifetime. Hence it is preferred to build temporary on-the-fly secure sessions in order to confirm intrusion whenever it is detected.

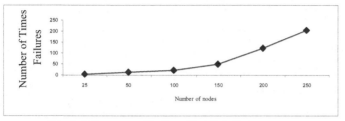

Fig. 5. Failure rate of Indirect Key Path Request

4.2 On-the-fly Secure Session Establishment schemes

Two simple on-the-fly schemes which are resistant to man-in-the-middle attack are proposed in this paper. The first scheme takes the help from the Base Station assuming nobody can compromise the Base Station. The second scheme uses multi-path communication. Since the keys generated for secure communication are used for only one session, their permanent storage in the memory is avoided.

4.2.1   Intrusion Confirmation protocol utilizing Base Station

The proposed scheme is a simple, energy efficient, low memory requirement scheme utilizing the Base Station and the pair-wise keys already established in the sensor networks between any node and the Base Station. These pair-wise keys are established using the protocol presented in our earlier work [7] in which a sensor node authenticates to the Base Station and establishes efficiently two pair-wise keys, one with Base Station and another with its cluster head.

The cluster head CH_k randomly generates a key K_{CHk-Si} and encrypts the key using the pair-wise key between CH_k and the Base Station.   The *Intrusion-Confirm* request message is prepared including the request type, encrypted K_{CHk-Si} and digital signature and the message is sent to the Base Station. The Base Station decrypts the pair-wise key, re-encrypts it with the pair-wise key shared with S_i and forwards the request message to S_i. Once the request message reaches S_i, it decrypts the pair-wise key and prepares the *Intrusion-Confirm* reply message. The reply includes the actual value supplied by the node and its parent ID. The reply is encrypted using K_{CHk-Si} and directly sent to CH_k, which decrypts the reply and stores the reply for further analysis. The *Intrusion-Confirm* request/reply protocol will continue with the parent of S_i until the replies are obtained from all the nodes in the aggregation tree path from S_i to CH_{k+1}. The protocol is given in Table 3. The digital signature generation and verification are included to preserve integrity and origin authentication. They use Elliptic curved based algorithm and the public keys which would have already been distributed as in our earlier work [12]. ECDSAG and ECDSAV denote the generation and verification of digital signature.

Table 3. Intrusion Confirmation through the Base Station

1	Cluster head CH_k when suspecting intrusion by CH_{k+1} on the data supplied from S_i, prepares *Intrusion-Confirm* request message.

- $K_{CHk-Si} = Genkey$ (); $(r, s) = ECDSAG$ (Request-type $\| S_i \| K_{CHk-Si}, Pr_{CHk}$)
- $CH_k \rightarrow BS$: Request-type $\| S_i \| E_{KCHk-BS}(K_{CHk-Si}) \| (r, s)$

2 When *BS* receives the *Intrusion-Confirm* request from CH_k

- Verifies digital signature by *ECDSAV* (Request-type $\| S_i\| K_{CHk-Si}, r, s, Pu_{CHk}$)
- $(r, s) = ECDSAG$ (Request-type $\| CH_k \| K_{CHk-Si}, Pr_{BS}$)
- $BS \rightarrow S_i$: Request-type $\| CH_k \| E_{KBS-Si}(K_{CHk-Si}) \| (r, s)$

3 When S_i receives the *Intrusion-Confirm* request from *BS*, prepares *Intrusion-Confirm* reply message.

- Verifies digital signature by *ECDSAV* (Request-type $\|CH_k \| K_{CHk-Si}, r, s, Pu_{BS}$)
- $S_i \rightarrow CH_k$: Reply-type, $E_{KCHk-Si}$(Data $\|$ *Cluster_ Head* (S_i))

4. When Cluster head CH_k receives the *Intrusion-Confirm* reply from S_i

- Decrypts the reply and stores (Data, S_i) for further analysis.
- Continue steps 1, 2 and 3 for *Cluster_Head* (S_i) until (*Cluster_ Head* (S_i) $\neq CH_k$)

4.2.2   Intrusion Confirmation protocol using Multi-path

In multi-path communication based scheme, CH_k finds out all possible routes between CH_k and S_i. The scheme utilizes the pair-wise keys already established in the sensor network between any cluster member and its cluster head by using the protocol presented in [7]. CH_k randomly generates a session key and prepares an *Intrusion-Confirm* request message to include the type of request and the encrypted session key. The request message is forwarded through multi-paths to S_i in a secure way by utilizing the pair-wise keys. Using the session key the source node encrypts the Intrusion-Confirm reply and returns the replies back to cluster head CH_k by using multi-paths.

For example let one such path be: CH_1 - CH_2 - CH_3 - S_4. CH_4 randomly generates the session key K_{CH1-S4} to be established, encrypts it with the pair-wise key between CH_1 - CH_2 and forwards the request message to CH_2. It decrypts the session key and re-encrypts it with the pair-wise key between CH_2 - CH_3. Finally the request will be reaching S_4. It decrypts the session key, prepares the reply by encrypting with K_{CH1-S4}. Once all the *Intrusion-Confirm* request messages from different routes arrive at S_4, it sends a common reply message to CH_1 through separate paths using multi-path communication. When CH_1 receives the reply, decrypts it and stores the reply. A compromised node in one of the routes may try to influence the reply. Therefore that reply may be different from all the other replies forwarded through other routes. By proper analysis, the requested cluster head can conclude the occurrence of intrusion.

4.2.3   Comparison of Key Establishment Protocols

The basic characteristics of the three protocols used for key establishment towards intrusion confirmation proposed in this paper lead us to provide a comparison of the protocols as given in Table 4. It brings out the superiority of the on-the-fly schemes in general and the on-the-fly scheme using the Base Station, in particular.

Paradoxically itself there are possibilities of a compromised cluster head executing the intrusion confirmation protocol many times to drain the battery power in the sensor nodes in the aggregation tree path. Only a scheme that involves the Base Station can detect this attack since the highly resourced Base Station can store the relevant Intrusion-Confirm request messages and can make an analysis to determine the event and establish the culprit. This attack is hard to detect in the permanent key scheme and in the multi-path based scheme and requires further study.

Table 4. Comparison of Key Establishment Protocols

Parameter	Post-deployment Permanent key scheme	On-the-fly scheme using	
		Base Station	Multi-path
Key Establishment	Initial phase	On-the-fly	On-the-fly
Security against man-in-the-middle attack	√	√	√
Security against multiple times execution of intrusion confirmation	X	√	X
Intervention of Base Station	X	√	X
Communication overhead for key establishment	High	Low	Depends on number of paths
Communication overhead for Intrusion Confirmation	Very low	Low	Medium
Memory for additional key storage	High	Nil	Nil
Speed of response after attack	Immediate	Slight Delay	Moderate Delay
Probability of failure in key establishment	>0	= 0	= 0
Energy expense	High	Low	Low

5. Intrusion Defense Scheme

We suggest a simple and practical intrusion defense scheme. The cluster head analyzes the *Intrusion-Confirm* Replies and if intrusion is confirmed, it sends an alarm message called *Intrusion-Alert* to the Base Station. The structure of the message is as shown in Table 5. When the Base Station receives the alert message, it increments the alert counter for this suspected node. When the alert counter value reaches a threshold, it verifies the validity of the intrusion directly with the suspected node because a group of compromised nodes can also supply false intrusion detection alerts about a legitimate node. As a second line of defense to verify the claim, the Base Station sends an encrypted request to the suspected node to return the latest aggregated value supplied from the cluster of S_i. If intrusion is confirmed, it decides to lower the level of trust to the compromised node and intimates the new trust level to all the neighbors of the compromised node. When a neighbor node receives the trust level message, it reduces the communication towards the compromised node.

Table 5. Structure of Intrusion-Alert Message

CH_k	CH_{k+1}	Data supplied by CH_{k+1}	S_i	Data supplied by S_i

6. Conclusion and Future Work

A new watchdog mechanism to detect node compromise attacks is proposed for a network that provides aggregation and security services. Also to detect false alarming, we require an intrusion confirmation protocol, which directly contacts all the nodes in the required aggregation tree path in a secure way and collects the details. We have proposed mainly two schemes. The first scheme is a post-deployment permanent key scheme, which uses polynomial distribution method to establish the required pair-wise keys. But this requires huge memory for key storage and high communication overhead. So we have proposed two simpler and energy efficient on the fly schemes: one by involving the Base Station and another by using multi-path communication. We have also suggested a feasible intrusion defense mechanism. The future work includes implementation and evaluation of the intrusion detection, confirmation analysis and defense schemes.

References:

1. Vijay Subhash Bhuse, Lightweight Intrusion Detection: A Second Line of Defense for Unguarded Wireless Sensor Networks, PhD Thesis, Western Michigan University, 2007.
2. Krontiris Ioannis, Tassos Dimitriou and Felix C. Freiling, Towards Intrusion Detection in Wireless Sensor Networks, Proceedings of 13th European Wireless Conference, 2007.
3. D. Liu and Peng Ning, Establishing Pair-wise Keys in Distributed Sensor Networks, Proceedings of ACM Conference on Computer and Communication Security, 2003.
4. Eschenauer and Gligor, A Key-management Scheme for Distributed Sensor Networks, Proceedings of ACM Conference on Computer and communication Security, 2004.
5. Chan, H., Perrig, A. and Song D., Random Key Pre-distribution Schemes for Sensor Networks, Proceedings of IEEE Symposium on Research in Security and Privacy, 2003.
6. Lingxuan Hu and David Evans, Secure Aggregation for Wireless Networks, Proceedings of Symposium on Applications and the Internet Workshops, pp. 384, January 2003.
7. Jibi Abraham and K S Ramanatha, An Efficient Protocol for Authentication and Initial Shared Key Establishment in Clustered Wireless Sensor Networks, Proceedings of 3rd IEEE/IFIP International Conference on Wireless and Optical Communications Networks, April 11-13, 2006, Bangalore, India.
8. Philip Levis, Nelson Lee, Matt Welsh and David Culler, TOSSIM: Accurate and Scalable Simulation of Entire TinyOS Applications, Proceedings of ACM Conference on Embedded Networked Sensor Systems, 2003, pp. 126-137.
9. Chris Karlof and David Wagner, Secure Routing in Wireless Sensor Networks: Attacks and Countermeasures, Proceedings of IEEE International Workshop on Sensor Network Protocols and Applications, May 2003, pp. 113-117.
10. Sasha Slijepcevic, Miodrag Potkonjak, Vlasios Tsiatsis, Scott Zimbeck and Mani B Srivatsa, On Communication Security in Wireless Ad-Hoc Sensor Networks, Proceedings of 11th IEEE International Workshops on Enabling Technologies, pp. 139 – 144, 2002.
11. Jason Hill, Robert Szewczyk, Alec et. All., System Architecture Directions for Network Sensors, Proceedings of International Conference on Architectural Support for Programming Languages and Operating Systems, 2000, pp. 93-104.
12. Jibi Abraham and K S Ramanatha, A Complete Set of Protocols for Distributed Key Management in Clustered Wireless Sensor Networks, Proceedings of IEEE International Conference on Multimedia and Ubiquitous Engineering, pp. 920-925, April 2007, Korea.

Proposal and Evaluation of a Rendezvous-based Adaptive Communication Protocol for Large-scale Wireless Sensor Networks

Mirai Wakabayashi[1] and Harumasa Tada[2] and Naoki Wakamiya[1] and Masayuki Murata[1] and Makoto Imase[1]

[1] Graduate School of Information Science and Technology, Osaka University, Osaka 565-0871 Japan
m-wakaba,wakamiya,murata,imase@ist.osaka-u.ac.jp
[2] Faculty of Education, Kyoto University of Education, Kyoto 612-8522 Japan
htada@kyokyo-u.ac.jp

Abstract. Recently, wireless sensor networks (WSNs) have attracted attentions of many researchers since they can be used for wide range of applications such as environmental monitoring, security, disaster prevention, environmental control in office buildings, and precision agriculture. Control mechanisms for WSNs should adapt to a variety of communication patterns which reflect application requirements and the situation. In this paper, we propose ARCP (Ant-based rendezvous communication protocol), a novel communication protocol for WSNs. ARCP is designed to be adaptive to a variety of communication patterns by taking the rendezvous-based approach, where sensor data are collected and delivered through nodes marked as rendezvous points. At the same time, ARCP acquires robustness to failures and scalability with respect to network size by adopting AntHocNet, which is an ad-hoc routing protocol inspired by foraging behavior of ants. Through simulation experiments, we show that ARCP outperforms existing communication protocols in adaptability, robustness, and scalability.

Keywords: Wireless Sensor Network, Data-centric Communication, Rendezvous-based Approach, Ant Routing

1 Introduction

Recently, wireless sensor networks (WSNs) have attracted attentions of many researchers [2, 1]. WSNs consist of a number of small sensing devices (*nodes*) with wireless communicating component and a base station as a sink of sensor data. WSNs can be used for a wide range of applications including environmental monitoring, environmental control in office buildings, precision agriculture, and so on [3].

A variety of communication patterns emerges in WSNs reflecting application requirements and the situation. For example, consider a WSN for habitat monitoring. A number of nodes are distributed over the monitored area. They collect environmental data, such as temperature, humidity, and wind direction. Then, they send the collected sensor data to a base station periodically. Once an animal is detected, some of nodes

Please use the following format when citing this chapter:

Wakabayashi, M., et al., 2008, in IFIP International Federation for Information Processing, Volume 264; Wireless Sensor and Actor Networks II; Ali Miri; (Boston: Springer), pp. 161–172.

begin to collect and send more detailed sensor data more frequently to track the behavior of the animal. Following the movement of the animal, nodes for detailed sensing change. In this way, the location and number of nodes involved in communication and their frequency dynamically change in accordance with the situation. Therefore, control mechanisms for WSNs must be adaptive to a variety of communication patterns and changes of the situation. Failures of nodes and links also occur for fragility of low-cost device and unstable and unreliable radio communication environment. For example, a node halts due to energy exhaustion or physical damages. Some obstacles cause radio interference, which prevents a node from exchanging messages with a physically neighboring node. Therefore, control mechanisms for WSNs must be robust to failures of nodes and links.

To accomplish robust, scalable, and energy-efficient communication in WSNs, several protocols has been proposed [15, 17, 16]. Directed diffusion is a data-centric communication paradigm [15, 10, 12], where messages are sent with the description of interested data, e.g. "send wind direction and speed when the temperature is higher than 25°C" or "notify if a fire has been detected", rather than the address of the destination node, e.g. "send sensor data to the node with address d". Directed diffusion has three variations of communication protocols: two-phase pull (TPP), one-phase pull (OPP), and push. In TPP or OPP, nodes which are intended to gather sensor data (called *data gathering nodes*) send the description of interested sensor data to the entire network using flooding and then nodes that can provide requested sensor data (called *data provision nodes*) respond. In TPP, data gathering nodes receive responses via multiple routes and choose the minimum delay route among them. In OPP, on the other hand, data gathering nodes receive all responses via the minimum delay route. In push, data provision nodes send samples of sensor data to the entire network using flooding and then data gathering nodes respond. Because of their mechanisms, TPP and OPP are appropriate when data provision nodes are more than data gathering nodes and push is appropriate when data gathering nodes are more than data provision nodes [9]. Therefore, in directed diffusion, an appropriate protocol must be selected a priori taking into account expected communication patterns or must be selected dynamically reflecting the situation by introducing some switching mechanism. In [17], a tree-based multicasting scheme for communication between a data provision node and multiple mobile data gathering nodes is proposed. In [16], a Steiner tree is constructed between data provision nodes and a data gathering node by using an ant colony algorithm for data-centric routing. Although they also enable robust, scalable, and energy-efficient communication, they can be applied only to one-to-many or many-to-one type of communication.

The rendezvous-based approach [13], which is a hybrid of pull and push, delivers sensor data indirectly via nodes marked as *rendezvous points (RPs)*. Both data provision nodes and data gathering nodes notify RPs of the description of sensor data they can provide or they are interested in (See Fig. 1). Data provision nodes deliver sensor data to RPs. Data gathering nodes retrieve sensor data from RPs. In the rendezvous-based approach, numbers, locations, and communication frequency of data provision nodes and data gathering nodes do not much affect the performance of communication. Locations of RPs, on the other hand, do affect the performance and thus RPs must be located appropriately according to the communication pattern and the situation. For example,

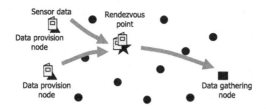

Fig. 1. Rendezvous-based approach

when there are more data provision nodes than data gathering nodes, an RP should be located closer to data provision nodes in order not to introduce much overhead in transmitting sensor data from data provision nodes to the RP. However, a mechanism to locate RPs has not been studied yet.

In this paper, we propose ARCP (Ant-based rendezvous communication protocol), a novel communication protocol for WSNs. Taking the rendezvous-based approach, ARCP is designed to be adaptive to a variety of communication patterns. ARCP appoints a node where delivery and retrieval of sensor data are expected to be frequent as an RP, promotes the usage of good RPs, and removes unused RPs. At the same time, ARCP acquires robustness to failures and scalability with respect to network size, by adopting AntHocNet [7], an ad-hoc routing protocol inspired by foraging behavior of ants. It is known that ants establish the shortest paths between the nest and food sources without centralized control by depositing volatile chemical substance called pheromone [8, 6, 5, 4].

This paper is organized as follows: Section 2 describes details of ARCP. Section 3 evaluates the adaptability, robustness, and scalability of ARCP through simulation experiments. Finally, Sect. 4 summarizes this paper and describes future works.

2 A Rendezvous-based Adaptive Communication Protocol for Large-scale Wireless Sensor Networks

In ARCP, nodes maintain neighborhood relation by periodically exchanging *hello messages*. Nodes send control messages called *routing ants* when they are ready to send or intend to receive sensor data. When routing ants from data provision nodes and those from data gathering nodes frequently encounter at a node, it is marked as an RP. Sensor data are transmitted by another kind of control messages called *carrier ants*. Carrier ants play similar role to data packets in AntHocNet. A *data provision carrier ant* generated by a data provision node carries sensor data from the data provision node to an RP. A *data gathering carrier ant* generated by a data gathering node carries sensor data from an RP to the data gathering node. Sensor data expires when a certain duration specified by the application passed from its generation. RPs discard expired sensor data. To improve delivery ratio of sensor data, carrier ants are guided to RPs with many arrivals of carrier ants, while unmarking RPs with few arrivals. All ants have the same TTL. When an ant travels more than $n_{TTL} > 0$ hops from its generation, it is discarded at the node.

Details of the behavior of ants in establishment and management of RPs and routes to them will be described in the following sections.

2.1 Behavior of routing ants

In ARCP, routes from data gathering nodes and data provision nodes to RPs are established and maintained by AntHocNet. In AntHocNet, routing information called *route pheromone* are maintained using three kinds of routing ants: reactive forward ants, proactive forward ants, and backward ants. Since ARCP is a data-centric communication protocol, a destination node is specified by the description of sensor data, rather than the address of the node.

Upon an application's request for delivery or retrieval of sensor data, a data provision node or a data gathering node s checks whether it has the pheromone for the description d. If the pheromone exists, the node s generates carrier ant and sends it at the regular intervals specified by the application. The carrier ant travels toward an RP choosing the next hop node according to pheromone values on each intermediate node. If no pheromone exists, the node s generates a routing ant. If the node s is a data provision node, the routing ant is a *data provision routing ant*. Otherwise, the routing ant is a *data gathering routing ant*. A routing ant generated at the node s behaves in the same way as a reactive forward ant in AntHocNet, which is sent to the next hop node chosen according to the pheromone for the description d or broadcast when there is no such pheromone. A node where data provision routing ants and data gathering routing ants frequently visit becomes an RP. When a routing ant marks a node as an RP or arrives at an RP, the routing ant becomes a backward ant in AntHocNet to return to the node s by traversing the same route it took to the RP. On the way to the node s, the routing ant updates pheromone values at the nodes along the route. Once the routing ant arrives at the node s, i.e. a route to the RP has been established, the node s begins to generate and send carrier ants.

In addition, as in AntHocNet, a data provision node and a data gathering node generate and send a data provision routing ant and a data gathering routing ant per n_r carrier ants, respectively, in order to maintain and improve routes. These routing ants behave as proactive forward ants of AntHocNet and they are sent to the next hop node chosen according to the pheromone.

2.2 Marking and Unmarking of RPs

Appropriate location of RPs depends on the number and position of data provision nodes and data gathering nodes. As mentioned in Sect. 2.1, nodes where data provision routing ants and data gathering routing ants frequently encounter are marked as RPs. It implies that a node is marked as an RP if it has many data provision nodes and data gathering nodes around it and has relayed many sensor data. By locating RPs closer to many data provision nodes and data gathering nodes, overhead in transmitting sensor data from data provision nodes to RPs and from RPs to data gathering nodes can be reduced. In this context, the term *encounter* means that a data gathering routing ant and a data provision routing ant pass the same node within a certain period of time.

Upon the arrival of a data provision routing ant at the node i, the timer X_p^i is set to the initial value $X_p > 0$. Likewise, upon the arrival of a data gathering routing ant at the node i, the timer X_g^i is set to the initial value $X_g > 0$. When the other timer is greater than zero on setting a timer, i.e. a certain period has not passed from the arrival of a routing ant from a node of the other type, the *encounter counter* C^i is increased by one. If the other timer is zero, the encounter counter C^i is decreased by one. If the encounter counter C^i of the node i reaches the value $n_{RP} > 0$, then the node i is marked as an RP. Since this marking is performed autonomously at each node, there is the possibility that multiple nodes are marked as an RP. Multiple RPs enable load distribution, where carrier ants are distributed among RPs. Consequently, the energy consumption at an RP and its neighbor will be suppressed and network congestion will be avoided. Robustness against RP failures will also be improved. However, because it is necessary to visit multiple RPs, not necessarily all though, to collect all sensor data, the gathering or delivery ratio of sensor data could decrease. The number of RPs can be reduced by, for example, using larger threshold n_{RP}. With larger n_{RP}, however, it takes long time for the first RP to appear. If the encounter counter C^i of the node i decreased to less than n_{RP}, i.e. few ants arrive at the node i, then the node i is unmarked. The unmarked node broadcasts a *link failure notification message* of AntHocNet in order to remove routes to itself.

2.3 Goodness of RPs

RPs are not equal in the frequency of arrival of carrier ants and the number of stored sensor data. The RP r is considered good when both data provision carrier ants and data gathering carrier ants arrive at r frequently and many sensor data are relayed at r. Guiding many ants to good RPs improves the delivery ratio of sensor data. For this purpose, in addition to the route pheromone, we introduce another kind of pheromone called *rendezvous pheromone* which reflects the goodness of the RP.

Since the route pheromone value T_{nd}^i in AntHocNet reflects only the latency of the route, ants prefer shorter routes rather than longer ones. In the rendezvous-based approach, however, the distance to an RP does not necessarily indicate the goodness of the RP. For example, consider the following case (See Fig. 2). There are two RPs r_1 and r_2 between data provision nodes and a data gathering node. The RP r_1 is near the data provision nodes, but it is far from the data gathering node. On the other hand, the RP r_2 is near the data gathering node, but it is far from the data provision nodes. In this case, while data provision carrier ants frequently visit the RP r_1, no data gathering carrier ant comes to the RP r_1, for choosing closer RP r_2. As a result, no sensor data can be delivered to the data gathering node via the RPs.

Now, we introduce rendezvous pheromone to tackle this problem. Ants choose the next hop node according to both the rendezvous pheromone value U_{nd}^i and the route pheromone value T_{nd}^i. On the way to the originating node, an ant at the node i updates the rendezvous pheromone value U_{nd}^i as well as the route pheromone value T_{nd}^i by the following equations:

$$T_{nd}^i = \gamma_T T_{nd}^i + (1 - \gamma_T) \tau_d^i \tag{1}$$

$$U_{nd}^i = \gamma_U U_{nd}^i + (1 - \gamma_U) n_{data} \tag{2}$$

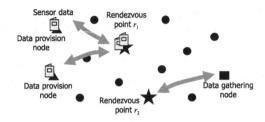

Fig. 2. RPs unshared by data provision node and data gathering node

Fig. 3. An example of network topology

where n is the node from which the node i received the ant, $\gamma_T \in [0,1]$ and $\gamma_U \in [0,1]$ are smoothing parameters, τ_d^i is the estimated time for a carrier ant to travel towards the RP (See [7] for details), and n_{data} is the number of sensor data stored at the RP when the ant left. The rendezvous pheromone value U_{nd}^i to an RP which has many sensor data becomes large by (2). An ant at the node i chooses the next hop node n for the description d with the probability P_{nd} given by the following equation:

$$P_{nd} = \frac{(T_{nd}^i)^{\beta_T}(U_{nd}^i)^{\beta_U}}{\sum_{j \in N_d^i}(T_{jd}^i)^{\beta_T}(U_{jd}^i)^{\beta_U}} \tag{3}$$

where N_d^i is the set of neighbors of the node i and $\beta_T, \beta_U > 0$ are parameters representing the weights of route pheromone and rendezvous pheromone respectively.

2.4   Transmission of Sensor Data

Data provision nodes deliver sensor data to RPs using data provision carrier ants. Data gathering nodes retrieve sensor data from RPs using data gathering carrier ants. Sensor data are associated with lifetime specified by an application. A data provision node holds its sensor data during their lifetime. An RP holds received sensor data during their lifetime, but it discards duplicated sensor data immediately. In addition, an RP aggregates sensor data [14], if an application requires. A data provision carrier ant carries all the sensor data stored at the data provision node to an RP. A data gathering carrier ant carries all the sensor data stored at an RP to the data gathering node. The sending rate of carrier ants at a data provision node and a data gathering node is determined according to an application requirement. A carrier ant travels toward an RP by choosing the

Table 1. Control parameter setting in simulation

β_T, β_U	Weight of pheromone	1 (routing ants) or 3 (carrier ants)
γ_T, γ_U	Smoothing parameter for pheromone	0.7
n_r	Number of carrier ant per data provision routing ant	30
n_{RP}	Threshold for encounter counter	10
t_{hello}	Hello message interval	30 s
X_p, X_g	Initial timer value for routing ants	240 s

next hop node with the probability given by (3) at each node and then returns to the originating node by traversing the same route it took to the RP. Note that carrier ants tend to choose better route than routing ants, because parameters β_T and β_U in (3) are set larger for carrier ants.

3 Simulation and Evaluation

In this section, we evaluate adaptability, robustness, and scalability of ARCP through simulation experiments.

3.1 Simulation Environment

We consider a WSN consisting of randomly placed immobile sensor nodes. A bidirectional link is established between any two nodes whose distance is less than 12 m. The link propagation delay is assumed to be 3 ms. It is assumed that no message is lost in the MAC layer due to e.g. collisions. Data provision nodes and data gathering nodes are chosen to form clusters [13]. First, one node is randomly chosen as a data provision node. Then, $n - 1$ nodes closest to the node are appointed as a data provision node. n data gathering nodes are chosen in the same way from the remaining nodes. Data provision nodes and data gathering nodes do not change during a simulation run. An example of network topology is shown in Fig. 3.

A data provision node generates sensor data every 2 s. For each of sensor data, the data provision node generates a data provision carrier ant. A data gathering node generates a data gathering carrier ant every 2 s. The size of single sensor data is set at 10 bytes. The size of an ant is set at 63 bytes except that a carrier ant amounts to $63 + 10k$ bytes where k is the number of sensor data it carries. The size of a hello message and a link failure notification message are set at 3 and 7 bytes respectively. The lifetime of sensor data is set at 40 s. Sensor data are not aggregated at RPs. Table 1 summarizes parameter setting in ARCP. Parameter setting for directed diffusion (TPP, OPP, and push) is based on [9].

As performance metrics, we use the delivery ratio of sensor data and the energy consumption. The delivery ratio for the whole network is defined as the average of the delivery ratio for all data gathering nodes. The delivery ratio for a data gathering node is defined as the ratio of the number of sensor data received by data gathering node to the number of sensor data generated by all data provision nodes. The energy consumption of the whole network is defined as the sum of the amount of energy consumed by

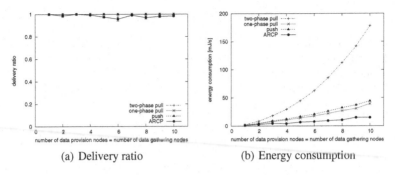

(a) Delivery ratio   (b) Energy consumption

Fig. 4. Adaptability to the number of data provision/gathering nodes

all nodes in sending and receiving messages. The energy consumption in sending and receiving a message is calculated according to [11]. Each simulation runs for 10,000 s in simulation time. We show the average and 95% confidence interval of results of five runs.

3.2   Evaluation of Adaptability

In order to evaluate the adaptability to communication patterns, we performed simulation experiments while changing the number of data provision nodes and data gathering nodes. 60 nodes are randomly placed in the $50 \times 50\,\mathrm{m}^2$ monitored area and the TTL of ants, n_{TTL} is set at 10.

Results are shown in Fig. 4. The number of data provision nodes which is equal to the number of data gathering nodes is changed from 1 to 10.

As shown in Fig. 4(a), the delivery ratio is always 1 in directed diffusion and is almost 1 in ARCP. That is, sensor data are successfully delivered to data gathering nodes regardless of the numbers of data provision nodes and data gathering nodes in any protocol. The reason for the slightly low delivery ratio of ARCP is that an ant chooses a next hop node in a probabilistic way for robustness against node failure. As such, an ant would take a long way and spend its TTL before reaching an RP. The delivery ratio can be improved by using larger n_{TTL}. However, it leads to larger energy consumption for allowing a longer route.

As shown in Fig. 4(b), ARCP is the most energy efficient protocol among the four under the many-to-many communication scenario. In ARCP, the increase of the number of data provision nodes and data gathering nodes only results in the increase of traffic between RPs and data provision nodes and between RPs and data gathering nodes. On the other hand, in directed diffusion, the increase of the number of data provision nodes and data gathering nodes results in the increase of the number of message flooding, which consumes the considerable amount of energy for involving all sensor nodes. Note that TPP costs the largest amount of energy, because data provision nodes first flood messages and then data gathering nodes send responses using all possible routes. In conclusion, ARCP is adaptive and scalable to the number of data provision nodes and data gathering nodes.

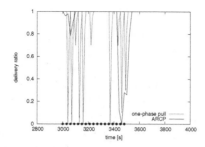

Fig. 5. Temporal variation of delivery ratio in a network with node failures

3.3   Evaluation of Robustness

In simulation experiments, 500 nodes are randomly arranged in the $140 \times 140\,\text{m}^2$ monitored area and n_{TTL} is set at 20. The numbers of data provision nodes and data gathering nodes are both set at 10. In order to evaluate robustness against node failures, we randomly choose 50 nodes among nodes which are not either of data provision nodes or data gathering nodes and stop them at 3,000 s. Failed nodes cannot send or receive any messages. All messages sent to failed nodes are lost. Any routing information and sensor data in failed nodes are removed. After 30 s, the failed nodes go back to normal operation and next 50 nodes are randomly chosen to halt. The same procedure is repeated until 3,500 s. For the comparison, we also consider OPP for its low energy consumption shown in Fig. 4(b).

Although not shown in figure, the average delivery ratio during the 500 s period is 77.5% in OPP and 80.5% in ARCP, respectively. In general, the rendezvous-based approach is vulnerable to failure of RP. If an RP fails and no other RP remains, the delivery ratio considerably degrades until a new RP is established. On the contrary to this conjecture, the delivery ratio of ARCP is higher than that of OPP.

To understand this result, we show an instance of temporal variation of average delivery ratio in Fig. 5. Circles on the horizontal axis in the figure indicate instances when node failures occurred. As can be seen, in OPP, the delivery ratio often drops to zero when node failure occurs. Although the delivery ratio is recovered soon by frequent message flooding, a large amount of sensor data is lost in this period. On the contrary, node failure does not affect the delivery ratio of ARCP in most cases, since ant-based routing allows ants to detour failed nodes by the probabilistic next hop selection. However, the delivery ratio of ARCP decreases to zero at about 3,430 s. At this time, an RP halted and there was no other RP. Due to autonomous and self-organizing behavior in establishing RPs and routes, it takes time to recover the delivery ratio once the only RP fails. However, the typical number of RPs was 2 in the simulation experiments and thus the probability that all RPs fail is low in the random failure scenario. Therefore, it can be concluded that ARCP is similarly to or slightly more robust than flooding-based deterministic protocol, i.e. OPP.

3.4 Evaluation of Scalability

In order to evaluate scalability with respect to the number of nodes, we conducted simulation experiments for 60, 500, 5,000, and 10,000 nodes. In the case of 60 nodes, they are distributed in the $50 \times 50\,m^2$ monitored area. We keep the density of nodes the same among the node population by changing the area. n_{TTL} is set in accordance with the diameter of a network as 10, 20, 70, and 100, respectively. The numbers of data provision nodes and data gathering nodes are both set as 10.

As shown in Fig. 6(a), the delivery ratio is kept 1 in OPP and push, since both use flooding to establish routes among data provision nodes and data gathering nodes. The delivery ratio of ARCP is also almost 1 for all of the node population, but we see the slight tendency of decrease as the number of nodes increases. As the population becomes large, ARCP takes more time to establish RPs and routes. Therefore, the amount of sensor data which expire at a data provision node before RP establishment increases. In addition, the length of routes also increases. Consequently, the probability that a carrier ant spends up its TTL becomes large, for taking a longer route by choosing a next hop node with small amount of pheromone. However, the delivery ratio is kept as high as 96%.

As shown in Fig. 6(b), the energy consumption increases in order of $O(N)$ in OPP and push and $O(\sqrt{N})$ in ARCP where N is the number of nodes, respectively. That is, ARCP is more scalable than others. Messages in OPP and push are classified into two categories: flooding and non-flooding. Since the number of flooding messages increases in proportional to the number of nodes, energy consumption increases in order of $O(N)$ for flooding messages. A non-flooding message is transmitted via the shortest route between a data provision node and a data gathering node. Since nodes are randomly located in the square monitored area, the length of the shortest route between two arbitrary nodes is in order of $O(\sqrt{N})$. Then, the amount of energy consumed by non-flooding messages is in order of $O(\sqrt{N})$. Since flooding is periodically performed, the energy consumption of OPP and push as a whole becomes in order of $O(N)$. On the other hand, in ARCP, it is difficult to estimate the energy consumption accurately, since ants sometimes broadcast themselves or choose a longer route. However, now, we approximate the energy consumption as follows. Most of ants are expected to choose the shortest route and their energy consumption is in order of $O(\sqrt{N})$. Some ants behave like a flooding message and their energy consumption is in order of $O(N)$. However, flooding behavior basically occurs to find an RP and establish a route, i.e. only at the initial stage. Therefore, we can approximate the energy consumption of ARCP as $O(\sqrt{N})$.

4 Conclusion

In this paper, we propose ARCP, a novel data-centric communication protocol for WSNs. ARCP combines the rendezvous-based approach and the ant-based routing protocol to be adaptive to communication patterns, robust to failures of nodes and links, and scalable with respect to network size. Through simulation experiments, we show that ARCP is more adaptive and scalable than existing communication protocols while keeping as high robustness as existing communication protocols.

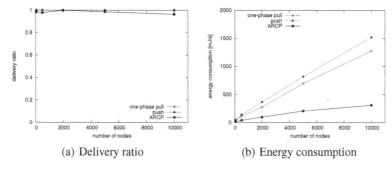

(a) Delivery ratio   (b) Energy consumption

Fig. 6. Scalability with respect to the number of nodes

We are considering further improvement on the adaptability of ARCP to dynamic changes in communication patterns. For this purpose, we need to develop a mechanism to dynamically move RPs to locations more appropriate for new condition.

Acknowledgments.   The authors would like to thank Associate Professor Hiroyuki Ohsaki of Osaka University for his fruitful suggestions. This work was partially supported by "The Global Center of Excellence Program" and a Grant-in-Aid for Scientific Research (A) (2) 16200003 from The Japanese Ministry of Education, Culture, Sports, Science and Technology.

References

1. Akkaya, K., Younis, M.: A survey on routing protocols for wireless sensor networks. Ad Hoc Networks **3**(3), 325–349 (2005)
2. Akyildiz, I., Su, W., Sankarasubramaniam, Y., Cayirci, E.: Wireless sensor networks: A survey. Computer Networks **38**(4), 393–422 (2002)
3. Arampatzis, T., Lygeros, J., Manesis, S.: A survey of applications of wireless sensors and wireless sensor networks. In: Proceedings of the 2005 IEEE International Symposium on, Mediterrean Conference on Control and Automation (ISIC05/MED05), pp. 719–724 (2005)
4. Beckers, R., Deneubourg, J.L., Goss, S.: Modulation of trail laying in the ant *Lasius niger* (Hymenoptera: Formicidae) and its role in the collective selection of a food source. Journal of Insect Behavior **6**(6), 751–759 (1993)
5. Beckers, R., Deneubourg, J.L., Goss, S., Pasteels, J.M.: Collective decision making through food recruitment. Insectes Sociaux **37**(3), 258–267 (1990)
6. Deneubourg, J.L., Aron, S., Goss, S., Pasteels, J.M.: The Self-Organizing Exploratory Pattern of the Argentine Ant. Journal of Insect Behavior **3**(2), 159–168 (1990)
7. Di Caro, G., Ducatelle, F., Gambardella, L.M.: AntHocNet: An adaptive nature-inspired algorithm for routing in mobile ad hoc networks. European Transactions on Telecommunications, Special Issue on Self-organization in Mobile Networking **16**(5), 443–455 (2005)
8. Goss, S., Aron, S., Deneubourg, J.L., Pasteels, J.M.: Self-organized shortcuts in the Argentine ant. Naturwissenschaften **76**, 579–581 (1989)
9. Heidemann, J., Silva, F., Estrin, D.: Matching data dissemination algorithms to application requirements. In: Proceedings of the 1st ACM conference on Embedded networked sensor systems (SenSys 2003), pp. 218–229 (2003)

10. Heidemann, J.S., Silva, F., Intanagonwiwat, C., Govindan, R., Estrin, D., Ganesan, D.: Building efficient wireless sensor networks with low-level naming. In: Proceedings of Symposium on Operating Systems Principles, pp. 146–159 (2001)
11. Heinzelman, W.R., Chandrakasan, A., Balakrishnan, H.: Energy-efficient communication protocol for wireless microsensor networks. In: Proceedings of the Hawaii International Conference on System Sciences (HICSS) (2000)
12. Krishnamachari, B., Estrin, D., Wicker, S.: Modelling data-centric routing in wireless sensor networks. Tech. Rep. CENG 02-14, USC Computer Engineering (2002)
13. Krishnamachari, B., Heidemann, J.: Application-specific modelling of information routing in wireless sensor networks. Tech. Rep. ISI-TR-576, USC-ISI (2003)
14. Rajagopalan, R., Varshney, P.K.: Data aggregation techniques in sensor networks: A survey. IEEE Communications Surveys and Tutorials 8(4), 48–63 (2006)
15. Silva, F., Heidemann, J., Govindan, R., Estrin, D.: Directed diffusion. Tech. Rep. ISI-TR-2004-586 (2004)
16. Singh, G., Das, S., Gosavi, S.V., Pujar, S.: Ant colony algorithms for steiner trees: An application to routing in sensor networks. Recent Developments in Biologically Inspired Computing pp. 181–204 (2004)
17. Zhang, W., Cao, G., Porta, T.L.: Dynamic proxy tree-based data dissemination schemes for wireless sensor networks. Wireless Networks 13(5), 583–595 (2007)

A Sensor Network Protocol for Automatic Meter Reading in an Apartment Building

Tetsuya Kawai[1] and Naoki Wakamiya[1] and Masayuki Murata[1] and Kentaro Yanagihara[2] and Masanori Nozaki[2] and Shigeru Fukunaga[2]

[1] Graduate School of Information Science and Technology, Osaka University, Japan
{t-kawai, wakamiya, murata}@ist.osaka-u.ac.jp
[2] Corporate R&D center, Oki Electric Industry Co., Ltd., Japan
{yanagihara726, nozaki765, fukunaga444}@oki.com

Abstract. A wireless sensor network for automatic meter reading needs to satisfy two contradicting requirements, i.e., long lifetime and prompt detection and notification of emergency. We propose a sensor network protocol for this purpose, in which sensor nodes operate on a low duty cycle while the latency of transmission is guaranteed to be less than a certain bound. In this protocol, each node is assigned a time slot in which it receives messages from other nodes. To accomplish slot assignment where nodes further from a BS are assigned earlier time slots for a packet to be transmitted to the BS in one cycle, we propose a slot assignment function with which a node can determine its own slot in a distributed way. We explore several slot assignment functions to find one which gives low and homogeneous contention over a grid network. The simulation results show that our protocol performs well close to the optimal case.

1 Introduction

Wireless sensor network (WSN) is a key technology to realize our safe, secure, and comfortable future life. Among its wide variety of applications, we focus our attention on automatic meter reading (AMR) in a large-scale apartment building which consists of hundreds of apartments. In such a building, meters are attached to pipes and cables in each apartment to monitor consumption of water, gas, electricity, and so on. A water, gas, or electricity company hires and sends personnel to collect meter readings once a month in most cases. In these days, those companies consider to adopt a WSN for meter reading, since it is costly to hire many personnel to cover all houses and buildings of customers and it becomes difficult for outsiders to enter modern apartment buildings for security reasons.

Such a WSN for AMR consists of meters equipped with a radio transceiver operated on battery power supply and a gateway server, i.e., base station (BS), connected to a monitoring station of a company through a regional wired or wireless network. Since a meter is usually stored in a meter box or pipe shaft, radio signal is heavily disturbed and the range of transmission is relatively small. Therefore, a WSN is sparse, where the average number of neighboring nodes, i.e., nodes in the range of radio signals, is a few. Monthly meter reading does not put too strict restriction on the delay requirement as far as meter reading is collected from all meters in a building. On the contrary, urgent

Please use the following format when citing this chapter:

Kawai, T., et al., 2008, in IFIP International Federation for Information Processing, Volume 264; Wireless Sensor and Actor Networks II; Ali Miri; (Boston: Springer), pp. 173–184.

information such as detection of gas leakage must be transmitted to the BS immediately, e.g., within 10 seconds, once it happens.

In this paper, we consider a network protocol for AMR, which can collect meter reading from all meters at a predetermined interval, e.g., once a month, and transmit urgent information from a meter to a BS within the specified delay bound. We assume that a contention-based MAC is adopted and that timers of nodes are synchronized by a certain synchronization protocol, such as proposed in [5, 6, 11]. In addition, a WSN is maintained by a certain topology control protocol, together with a health check mechanism conducted once a day, for example. Under these assumptions, we consider sleep scheduling for energy-efficient operation of a WSN.

The trade-off between latency and energy-efficiency has been discussed in some literatures [2–4, 9]. Sleep scheduling to reduce latency is also proposed in [1, 8, 10]. However, they do not consider reliability of data transmission. Both loss for unreliable wireless communication and latency caused by sleeping nodes are considered in the dynamic switch-based forwarding [7], which optimizes the expected delivery ratio, expected communication delay, or expected energy consumption. However, this forwarding method does not provide any guarantee on end-to-end delay. We aim to satisfy both requirements on bounded end-to-end delay and low duty cycle by waking up nodes from the edge of a WSN to the BS adopting a distributed and self-organizing slot assignment mechanism.

The rest of this paper is organized as follows. First we describe the details of our protocol in Sect. 2. In Sect. 3, we define the contention degree, which represents the intensity of contention at each node, and expected contention degree in a grid network is presented. Next we evaluate four slot assignment functions in Sect. 4 and the best one is further tuned for a grid network in Sect. 5. We conclude this paper in Sect. 6.

2 Sensor Network Protocol for AMR

To save energy consumption, making nodes sleep is one of the primary techniques. To detect an event and notify a BS of it, a node has to wake up at least once per the delay bound, denoted as d_{max} hereafter. Then, all nodes on the path from the detecting node to the BS must wake up at appropriate timing so that the urgent information is relayed to the BS immediately.

Now consider the case that the duration of one operation cycle, d_{max}, is divided into time slots of duration t_s. In most of TDMA-based scheduling, a slot is assigned to each node for packet transmission. A node can occupy the wireless channel during the assigned slot and send a packet without being disturbed by other nodes. This scheduling enables efficient usage of the wireless channel, but it is inefficient from an energy point of view. Since a node does not know which node has a packet to relay, it must listen to the wireless channel in all of slots assigned to its neighbors. In a WSN, such idle listening is the major drain of energy, especially when packets are generated intermittently as for AMR.

Our approach to shorten this idle listening is to assign time slots not for sending but for receiving to lower the duty clcle of nodes. With this slot assignment, each node has to be awake during only one slot per cycle even if it has two or more child nodes. In our sensor network protocol for automatic meter reading, called Sleep Scheduling by Distributed Slot Assignment (SSDSA), every node keeps awake for t_s at an interval of d_{max} for possible packet reception. The total number N of slots is given by $N = d_{max}/t_s$. We define slotID $0 \le k \le N - 1$ where a smaller slotID corresponds to an earlier slot in a cycle of d_{max}. The duration t_s is determined to be long enough for MAC layer to deliver a packet to a next-hop node, including carrier sense, MAC level acknowledgement, and retransmissions. For example, its typical value would be between 100 ms and 200 ms for IEEE 802.15.4 and smaller for IEEE 802.11. If t_s = 100 ms and d_{max} = 10 seconds, the duty cycle is 1/100. Even lower duty cycle can be achieved by employing an energy-aware low duty MAC protocol such as [12]. In SSDSA, a node further from a BS obtains an earlier slot with a smaller slotID. As far as every node has a next-hop node which has a slot of a larger slotID, a packet originating at any node can reach the BS within d_{max}.

The details of SSDSA are as follows:

1. New node i first tries to discover neighbor nodes by a neighbor discovery protocol employed. Neighbor discovery is out of scope of this paper.

2. When node i discovers neighbor j, node j notifies node i of its ID, level l_j, which corresponds to the hop distance from the BS, and slotID k_j. We assume that the BS has a power supply and does not sleep. The BS declares slotID N and the level 0. Node i stores the received information in a neighbor table. Neighbors are listed in ascending order of the level as the first key and the slotID as the second key (see Fig. 1). A time synchronization process could be conducted in this stage. In the rest of this paper, we refer a parent or child node as a direct neighbor of a node which is one hop closer or further to the BS than the node itself, respectively. A next-hop node is one of parent nodes to which a node sends a packet and a preceding node is one of child nodes from which a packet is received.

3. After completing the neighbor discovery process, node i determines its own level l_i and slotID k_i as $l_i = l_i^1 + 1$ and $k_i = f(k_i^1)$, where l_i^1 and k_i^1 are the level and slotID of the node at the top of the neighbor table, respectively. $f(k)$ is called the slot assignment function (SAF), which gives a smaller value than k following the slot assignment probability distribution function (SAPDF), which will be discussed later. Level $l_i + 1$ nodes and level l_i nodes with slotID equal or smaller than the maximum among level $l_i - 1$ nodes are removed from the neighbor table. In Fig. 1, node B does not have node C in its neighbor table for this reason. Then, node i sends these information to its neighbor nodes of the same level at their slots. A neighbor node which receives this notification adds node i to its neighbor table, if slotID k_i is greater than the maximum slotID of its parent nodes.

4. Once node i determines its time slot k_i, node i wakes up at the slot, keeps active for t_s, and goes back to sleep.

5. With the slot assignment stated above, a node can have two or more next-hop candidates for improving reliability of transmission [7]. If node i receives a packet during its time slot or it has a packet to send, it wakes up again at the next-hop node's time slot, i.e, k_i^1, and sends the packet. If node i fails in transmitting a packet to the first next-hop node, it tries the second next-hop node in the neighbor table at slot k_i^2 (see node C in Fig. 2). If the transmission fails again, it tries third one. It repeats this

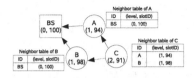
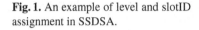

Fig. 1. An example of level and slotID assignment in SSDSA.

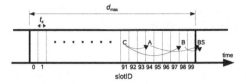

Fig. 2. Scheduled packet transmission.

procedure until the transmission succeeds or fails at the last next-hop node in the neighbor table.

Possible congestion around the BS could be avoided, for example, by giving level 1 nodes even slotIDs and having them transmit a packet at the following time slot of an odd slotID. Note that, in SSDSA, sleep scheduling of nodes and routing are integrated in the time slot assignment. Slot assignment is done in a fully-distributed and self-organizing manner by using neighbor information.

3   Contention Degree

3.1   Definition

In SSDSA, it is possible that two or more nodes transmit packets at the same time, since time slots are assigned not for transmission but for reception. Moreover, since each node determines its own slotID in a distributed manner, two or more nodes could have the same slotID. These could degrade the reliability of transmission and lead to extra energy consumption in the MAC layer. Therefore, we need to balance the degree of contention among nodes.

We define the contention degree C of a node as the number of neighbors which can transmit a packet at the time slot given to the node. For example, in Fig. 3, the contention degree of node A is three, where neighboring nodes B, C, and D compete for slot 80. Although we consider only the first next-hop node to compute the contention degree in the following discussion, it is straightforward to extend this idea to involve the second and more next-hop nodes by weighing contribution to the contention degree, 1.0 for the first next-hop node and 0.1 for the second next-hop node for example.

3.2   Contention Degree in a Grid Network

First, we calculate the expected contention degree in a grid network, considering the arrangement of apartments in a building. In a grid network shown in Fig. 4, each circle represents a node and a number inside denotes its level. Each line corresponds to a bidirectional link among nodes. The contention degree C_B of node B having two child nodes X and Y is 0, 1, or 2, depending on slotIDs of nodes A, B, and C. For example, for $k_A > k_B > k_C$, $C_B = 1$, since node X chooses node B and node Y chooses node C as their first next-hop node.

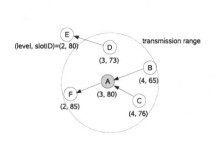

Fig. 3. An example of the contention degree.

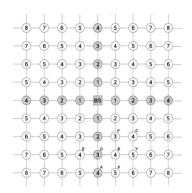

Fig. 4. A grid network.

Now let q the probability that two nodes of level l have the same slotID. The probability that two level l nodes have different slotIDs becomes $(1-q)/2$. Then, $P(k_A > k_B > k_C) = \{(1-q)/2\}^2$. Calculating the probability of all nine combinations of k_A, k_B, and k_C, the expected contention degree of node B is given as $E(C_B) = 1+q$. Next we consider nodes represented by a gray circle in Fig. 4. They have three child nodes and one parent node. For example, node D has child nodes A, B, and E. We can apply similar discussion as above for nodes B and E. On the other hand, node A always chooses node D as its next-hop node. Therefore, the expected contention degree of node D is given as $E(C_D) = 2+q$. We refer a gray node as "m-node" (main stream node) and an open-circle node as "b-node" (branch stream node), and we define the expected contention degree of m-nodes, $E(C_m)$, and of b-nodes, $E(C_b)$, as

$$E(C_m) = 2+q, \ E(C_b) = 1+q. \tag{1}$$

4 Slot Assignment Probability Distribution Functions

4.1 Evaluation Metrics

In the rest of this paper, we explore SAFs which lead to lower and more homogeneous contention degree. We employ following four evaluation metrics.

- $E_l(C)$. The average contention degree of level l nodes.
- $V_l(C)$. The variance of the contention degree of level l nodes.
- p_{empty}. The ratio of unassigned time slots.
- q_{isolated}. When all of parent nodes are assigned the time slot 0, a node cannot obtain its time slot and it is isolated. q_{isolated} is the ratio of isolated nodes.

4.2 Examples of SAPDF

In this subsection, we consider SAPDF $g^k(x)$ which determines SAF $f(k)$. $g^k(x)$ gives the probability that slotID x is assigned to a node when the slotID of its first next-hop node is k. Letting $F_l(x)$ denote the probability distribution of nodes (PDN) that a node of level l is assigned slotID x, we have

$$F_l(x) = \frac{1}{n_l} \sum_{i \in \mathscr{S}_l^*} g^{k_i}(x), \tag{2}$$

where n_l and $\mathscr{S}_l^*$ are the number of level l nodes and a set of their first next-hop nodes, respectively. Assuming the distribution of slotIDs of nodes in $\mathscr{S}_l^*$ is identical to that of all level $l-1$ nodes, $F_{l-1}(x)$, (2) can be rewritten as

$$F_l(x) = \sum_{k=0}^{N-1} g^k(x) F_{l-1}(k). \tag{3}$$

The duplication probability q in (1) can be derived as

$$q = \sum_{x=0}^{N-1} \{g^k(x)\}^2. \tag{4}$$

We consider the following four typical examples of SAPDFs.

K-1 If the slotID of the first next-hop node is k, slotID $k-1$ is assigned.

$$g^k(x) = \begin{cases} 1 \ (x = k-1) \\ 0 \ (0 \le x \le k-2, k \le x \le N-1). \end{cases} \tag{5}$$

According to K-1, all level l nodes have the identical slotID $N-l$. Therefore, K-1 leads to the worst case scenario.

L-BOUND The lower bound L_l of slotID for level l nodes is predetermined. A slotID is randomly chosen between L_l and $k-1$.

$$g^k(x) = \begin{cases} \frac{1}{k-L_l} \ (L_l \le x \le k-1) \\ 0 \ (0 \le x \le L_l-1, k \le x \le N-1). \end{cases} \tag{6}$$

LINEAR A slotID is randomly chosen between 0 and $k-1$ according to linear distribution as shown in Fig. 5(a).

$$g^k(x) = \begin{cases} \frac{2}{k(k+1)}(x+1) \ (0 \le x \le k-1) \\ 0 \qquad\qquad (k \le x \le N-1). \end{cases} \tag{7}$$

PDN $F_l(k)$ of LINEAR for level 1 through 3 calculated by (3) and (7) is shown in Fig. 5(b).

EXPONENTIAL A slotID is randomly chosen between 0 and $k-1$ following the exponential distribution as shown in Fig. 6.

$$g^k(x) = \begin{cases} e^{-\lambda_k\{k-(x+1)\}} - e^{-\lambda_k(k-x)} \ (k \ne 1, 0 \le x \le k-1) \\ 0 \qquad\qquad\qquad (k \ne 1, k \le x \le N-1). \end{cases} \tag{8}$$

SlotID 0 is assigned for $k=1$.

Now consider the probability q for $k=100$. In K-1, all nodes of the same level have the same slotID, thus $q=1$. In L-BOUND, q heavily depends on L_l and is given as $1/(100-L_l)$. In LINEAR, q becomes 0.013. For EXPONENTIAL, when we set λ_k to $11.5/(k-1)$ to have the ratio of isolated nodes at level 10 less than 0.01 % following the discussion in Sect. 4.3, q becomes 0.058.

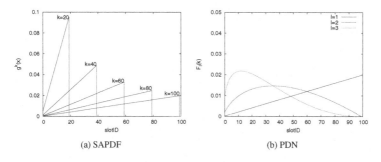

(a) SAPDF

(b) PDN

Fig. 5. (a) SAPDF and (b) PDN of LINEAR.

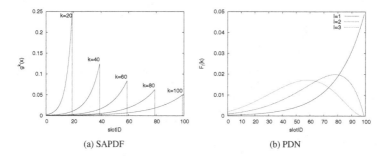

(a) SAPDF

(b) PDN

Fig. 6. (a) SAPDF and (b) PDN of EXPONENTIAL ($\lambda_k = 5/(k-1)$).

4.3 Parameter Determination of EXPONENTIAL SAPDF

In EXPONENTIAL, parameter λ_k determines the distribution of slotIDs. With small λ_k, slotIDs are widely distributed, with which probability q becomes small and the contention degree becomes low. However, as an adverse effect, it generates many isolated nodes by assigning a small slotID to a low level node. Therefore, we need to find appropriate λ_k which guarantees the maximum ratio of isolated nodes of level l at the desired value $1 - P^\star$. We define P_l as the probability that slotID $k = [1, N-1]$ is assigned to a level l node, namely,

$$P_l = \sum_{i=1}^{N-1} F_l(i) = 1 - F_l(0). \tag{9}$$

Among level 2 nodes, those who have a level 1 node with slotID $k = 0$ as their next-hop node cannot obtain a valid slotID and they are isolated. For simplicity, ignoring a case that a level 2 node has two or more parent nodes, $1 - P_1$ of level 2 nodes are isolated. Among the rest, P_2 get slotID $k = [1, N-1]$ and $1 - P_2$ get slotID $k = 0$. Therefore, among all level 2 nodes, the probability of getting slotID $k = [1, N-1]$ and $k = 0$ are $P_1 P_2$ and $P_1(1 - P_2)$ respectively. Assuming $P = P_1 = P_2 = \cdots = P_l$ for simplicity, the ratio of level l nodes with a valid slotID is P^{l-1}. Since this must be larger than $P^\star$, we have

$$P > \sqrt[l-1]{P^\star}. \tag{10}$$

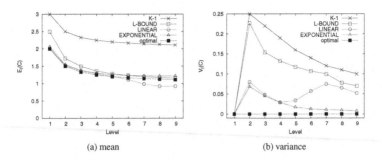

(a) mean   (b) variance

Fig. 7. The contention degree.

From (3) and (9), we get

$$P = \sum_{i=1}^{N-1} F_l(i) = \sum_{i=1}^{N-1} \left(\sum_{k=0}^{N-1} g^k(i) F_{l-1}(k) \right) = \sum_{k=0}^{N-1} \left(F_{l-1}(k) \sum_{i=1}^{N-1} g^k(i) \right). \quad (11)$$

If

$$\sum_{i=1}^{N-1} g^k(i) > \sqrt[l-1]{P^\star} \quad (12)$$

holds,

$$P = \sum_{k=0}^{N-1} \left(F_{l-1}(k) \sum_{i=1}^{N-1} g^k(i) \right) > \sqrt[l-1]{P^\star} \sum_{k=0}^{N-1} F_{l-1}(k) = \sqrt[l-1]{P^\star}. \quad (13)$$

Therefore, (12) is a sufficient condition for (10). For EXPONENTIAL, from (8),

$$\sum_{i=1}^{N-1} g^k(i) = 1 - e^{-\lambda_k(k-1)}. \quad (14)$$

Substituting (14) into (12), we finally get

$$\lambda_k > -\frac{\log(1 - \sqrt[l-1]{P^\star})}{k-1}. \quad (15)$$

4.4 Simulation Experiments

We compare the four SAPDFs by simulation experiments. 220 nodes up to level 10 are arranged in a grid network centered at a BS. The number of time slots N is set to 100. For L-BOUND, the lower bound L_l is determined according to the number of nodes belonging to level l, $(L_{l-1} - L_l)/(L_l - L_{l+1}) = l/(l+1)$ where $L_0 = N$ and $L_{10} = 0$. For EXPONENTIAL, $\lambda_k = 11.5/(k-1)$. Shown results are averaged over 500 simulation runs.

In the optimal case, the contention degree of all level l nodes should be equal to the average number of level $l+1$ nodes per node, thus

$$E_l^{\text{opt}}(C) = (l+1)/l, \quad V_l^{\text{opt}}(C) = 0. \quad (16)$$

Table 1. Evaluation metrics for four SAPDFs.

| | K-1 | L-BOUND | LINEAR | | EXPONENTIAL | | |
					$r = 2.0$	3.0	4.0
p_{empty} (%)	90.0	11.8	65.9	29.4	31.9	33.7	35.1
$q_{isolated}$ (%)	0	0	41.0	0.008	0.006	0	0

The results are shown in Fig. 7 and Table 1. As shown in Fig. 7(a), K-1 gives the largest contention degree, whereas there is no isolated node as shown in Table 1. L-BOUND can efficiently assign slots, but the contention degree and its variance are not sufficiently small. LINEAR accomplishes the smallest contention degree at high level nodes. However, it is due to isolated nodes. In Fig. 5(b), the peak of PDN for level 3 is at $k = 12$, which means that higher level nodes have little choice of slotIDs. Consequently, 41 % of nodes are isolated and do not contend for the wireless channel. We can see that the graph of EXPONENTIAL is very close to that of the optimal and the best among four SAPDFs.

The variance $V_l(C)$ of L-BOUND is higher than that of LINEAR or EXPONEN-TIAL as shown in Fig 7(b). Basically, an m-node has larger contention degree than a b-node as indicated by (1) if they are equally chosen as a next-hop node. However, at higher levels in LINEAR or EXPONENTIAL, a b-node has a smaller slotID than an m-node for having more parent nodes and it is chosen as a next-hop node more often. Then, the difference of the contention degree between m-nodes and b-nodes becomes smaller. On the contrary, in L-BOUND, a node randomly selects a slotID within a pre-determined range of time slots. EXPONENTIAL gives the smallest variance among the four SAPDFs.

5 Optimization of Exponential Distribution

As shown in the previous section, EXPONENTIAL gives the near-optimal performance. However, m-nodes have higher contention degree than b-nodes for having more child nodes especially at lower levels. We consider to make the contention degree same or similar among m-nodes and b-nodes by using different parameters.

5.1 Stochastic Analysis

The basic idea is to make a node which has both m-nodes and b-nodes as its parents choose a b-node as its next-hop node with a higher probability. For this purpose, we employ a larger coefficient of the exponential distribution for m-nodes to get a larger slotID than those of b-nodes.

In a grid network shown in Fig. 4, introducing probability p that an m-node has a larger slotID than that of a b-node of the same level, probability q_1 that an m-node has an equal slotID to that of a b-node, and probability q_2 that two b-nodes have the identical slotID, we obtain the expected contention degree $E(C_A)$, $E(C_B)$, and $E(C_C)$ of m-node A, b-node B which has an m-node of the same level as its two-hop neighbor, and

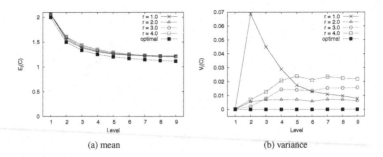

Fig. 8. The contention degree with the parameter tuning.

b-node C which does not have any m-nodes of the same level as its two-hop neighbors, respectively. After similar calculation as in Sect. 3.2, we get

$$E(C_A) = 3 - 2p, \ E(C_B) = p + q_1 + \frac{1 + q_2}{2}, \ E(C_C) = 1 + q_2. \tag{17}$$

For $E(C_A) = E(C_B) = E(C_C)$, erasing q_2, we obtain

$$2p + q_1 = \frac{3}{2}. \tag{18}$$

This is a necessary condition for the identical contention degree among nodes of the same level. Now, let us assume that an m-node and a b-node have their next-hop node with slot k and their slotIDs are assigned by SAPDFs g_m^k and g_b^k respectively. The probabilities p and q_1 can be calculated as

$$p = \sum_{x=1}^{N-1} \left(g_m^k(x) \sum_{i=0}^{x-1} g_b^k(i) \right), \ q_1 = \sum_{x=0}^{N-1} g_m^k(x) g_b^k(x). \tag{19}$$

In order to assign a larger slotID to an m-node, we employ a larger coefficient for g_m^k than g_b^k by introducing a parameter $r \geq 1$ as,

$$g_m^k(x) = \begin{cases} e^{-r\lambda_k\{k-(x+1)\}} - e^{-r\lambda_k(k-x)} & (0 \leq x \leq k-1) \\ 0 & (k \leq x \leq N-1) \end{cases} \tag{20}$$

while $g_b^k(x)$ is the same as (8). Using (19) and (20), we can derive the optimal r to have (18), although results are not shown for space limitation.

5.2 Simulation Experiments

The mean $E_l(C)$ and variance $V_l(C)$ of the contention degree at level l are plotted in Fig. 8 by changing r. $E_l(C)$ slightly increases with r, because q_1 in (17) increases. $V_2(C)$ dramatically decreases by changing r compared to the case without the parameter tuning ($r = 1.0$). It means that the contention degree is well equalized among m-nodes and b-nodes. At level 3 or above, $r = 2.0$ gives better results than $r = 3.0$ does. As mentioned in Sect. 4.4, even in the case without the parameter tuning, slotIDs of b-nodes tend to be smaller than those of m-nodes as the level increases. Increasing r enhances this effect even further, thus p becomes too large to satisfy (18) with $r \geq 3.0$ at higher levels. Table 1 shows p_{empty} and $q_{isolated}$ for each of four r values. As r increases, p_{empty} also increases, since m-nodes are distributed within a narrower range of time slots, in other words, more packed.

6 Conclusion

In this paper, a WSN protocol for AMR in a large-scale apartment building was proposed. For accomplishing both a low duty cycle and delay-bounded transmission, the operation interval is set at the delay bound and divided into time slots. A node determines its time slot for packet reception in a distributed and stochastic manner, so that every node has a next-hop node with a larger slotID. It was shown that EXPONENTIAL SAPDF enabled less and more identical contention degree at all level nodes in a grid network and the results were near-optimal. Our future work includes evaluation of energy efficiency of our protocol in practical experiments, including comparison to a TDMA-type protocol, which we will report in near future.

Acknowledgement This research was supported in part by "Global COE (Centers of Excellence) Program" of the Ministry of Education, Culture, Sports, Science and Technology, Japan.

References

1. Cao, Q., Abdelzaher, T., He, T., Stankovic, J.: Towards optimal sleep scheduling in sensor networks for rare-event detection. In: Proceedings of the 4th International Symposium on Information Processing in Sensor Networks (IPSN 2005), pp. 20–27. Los Angeles, California, USA (2005)
2. Chiasserini, C.F., Garetto, M.: Modeling the performance of wireless sensor networks. In: Proceedings of 23rd Annual Joint Conference of the IEEE Computer and Communication Societies (INFOCOM 2004), pp. 220–231. Hong Kong (2004)
3. Cohen, R., Kapchits, B.: An optimal algorithm for minimizing energy consumption while limiting maximum delay in a mesh sensor network. In: Proceedings of the 26rd Anuual Joint Conference of the IEEE Computer and Communications Societies (INFOCOM 2007), pp. 258–266. Anchorage, Alaska, USA (2007)
4. Dousse, O., Mannersalo, P., Thiran, P.: Latency of wireless sensor networks with uncoordinated power saving mechanisms. In: Proceedings of the 5th ACM international symposium on Mobile ad hoc networking and computing (MobiHoc '04), pp. 109–120. Tokyo, Japan (2004)
5. Elson, J., Girod, L., Estrin, D.: Fine-grained network time synchronization using reference broadcasts. In: Proceedings of the 5th symposium on Operating systems design and implementation (OSDI '02), pp. 147–163. Boston, Massachusetts, USA (2002)

6. Ganeriwal, S., Kumar, R., Srivastava, M.B.: Timing-sync protocol for sensor networks. In: Proceedings of the 1st international conference on Embedded networked sensor systems (SenSys '03), pp. 138–149. Los Angeles, California, USA (2003)

7. Gu, Y., He, T.: Data forwarding in extremely low duty-cycle sensor networks with unreliable communication links. In: Proceedings of the 5th international conference on Embedded networked sensor systems (SenSys '07), pp. 321–334. Sydney, Australia (2007)

8. Keshavarzian, A., Lee, H., Venkatraman, L.: Wakeup scheduling in wireless sensor networks. In: Proceedings of the 7th ACM international symposium on Mobile ad hoc networking and computing (MobiHoc '06), pp. 322–333. Florence, Italy (2006)

9. Lai, W., Paschalidis, I.C.: Sensor network minimal energy routing with latency guarantees. In: Proceedings of the 8th ACM international symposium on Mobile ad hoc networking and computing (MobiHoc '07), pp. 199–208. Montreal, Quebec, Canada (2007)

10. Lu, G., Sadagopan, N., Krishnamachari, B., Goel, A.: Delay efficient sleep scheduling in wireless sensor networks. In: Proceedings of the 24th Annual Joint Conference of the IEEE Computer and Communications Societies (INFOCOM 2005), pp. 2470–2481. Miami, Florida, USA (2005)

11. Maróti, M., Kusy, B., Simon, G., Ákos Lédeczi: The flooding time synchronization protocol. In: Proceedings of the 2nd international conference on Embedded networked sensor systems (SenSys '04), pp. 39–49. Baltimore, MD, USA (2004)

12. Ye, W., Silva, F., Heidemann, J.: Ultra-low duty cycle mac with scheduled channel polling. In: Proceedings of the 4th international conference on Embedded networked sensor systems (SenSys '06), pp. 321–334. Boulder, Colorado, USA (2006)

Monitoring Linear Infrastructures Using Wireless Sensor Networks *

Imad Jawhar, Nader Mohamed, Khaled Shuaib and Nader Kesserwan

College of Information Technology
United Arab Emirates University
P.O. Box 17551, Al Ain, UAE
E-mail: {ijawhar, nader.m, k.shuaib, nkesserwan}@uaeu.ac.ae

Abstract. This paper presents and evaluates a protocol for Linear Structure wireless sensor networks which uses a hierarchical addressing scheme designed for this type of networking environment. This kind of linear structure exists in many sensor applications such as monitoring of international borders, roads, rivers, as well as oil, gas, and water pipeline infrastructures. The networking framework and associated protocols are optimized to take advantage of the linear nature of the network to decrease installation, maintenance cost, and energy requirements, in addition to increasing reliability and improving communication efficiency. In addition, this paper identifies some special issues and characteristics that are specifically related to this kinds of networks. Simulation experiments using the proposed model, addressing scheme and routing protocol were conducted to test and evaluate the network performance under various network conditions.

Keywords: Ad hoc and sensor networks, routing, addressing schemes, wireless networks.

1 Introduction

The advent of technology in computing and electronics is pioneering an emerging field of tiny wireless sensors, offering an unprecedented opportunity for a wide array of real time applications. In recent years, wireless sensor networks are emerging as a suitable new tool for a spectrum of new applications [1]. These tiny sensor nodes are low cost, low power, easily deployed, and self-organizing. They are usually capable of local processing. Each sensor node is capable of only a limited amount of processing, but when coordinated with the information from a large number of other nodes, they have the ability to measure a given physical environment in great detail.

Research in the field of Wireless Sensor Networks is relatively active and involves a number of issues that are being investigated. These issues are efficient routing protocols for ad hoc and wireless sensor networks [8], QoS support [7][9], security [2], and middleware [4]. Most of these issues are investigated under the assumption that the network used for sensors does not have a predetermined infrastructure [3][5][10][11]. Fortunately, the wireless sensor network needed for monitoring linear infrastructures

---

* This work was supported in part by UAEU Research grant 08-03-9-11/07.

Please use the following format when citing this chapter:

Jawhar, I., et al., 2008, in IFIP International Federation for Information Processing, Volume 264; Wireless Sensor and Actor Networks II; Ali Miri; (Boston: Springer), pp. 185–196.

will be a structured network in which all sensor nodes will be distributed in a line. This characteristic can be utilized for enhancing the communication quality and reliability in this kind of networks.

This paper addresses the issues and challenges of using wireless sensor networks that are aligned in a linear formation for monitoring and protection of critical infrastructures and geographic areas. Also, it presents a routing protocol and addressing scheme for this special kind of sensor networks. As mentioned earlier, this kind of alignment of sensors can arise in many applications such as the monitoring and surveillance of international boundaries for illegal crossing, or smuggling activities, monitoring of roads, or long pipelines carrying oil, gas and water resources, river environmental monitoring, as well as many other such uses. The presented architecture utilizes the special linear structure of the networks to solve some of communication reliability and security problems. The objective of the design is to reduce installation and maintenance costs, increase network reliability and fault tolerance, increase battery life for wireless sensors, reduce end-to-end communication delay for quality of service (QoS) sensitive data, and increase network lifetime by utilizing the special linear structure of the network. This paper extends the model and architecture discussed in [6]. More details on the background, motivation, advantages, and applications for using linear structure wireless sensor networks can be found in that paper.

There are many advantages for using wireless sensor network technology to provide protection and monitoring of linear infrastructures such as oil, water, and gas pipelines, international borders, roads and rivers. Some of these advantages are: (1) Faster and less costly network deployment. (2) Additional savings in network maintenance and necessary personnel expertise. (3) Increased reliability and security due to the ability to disseminate collected information at designated wireless access points, and the ability to introduce flexible multihop routing which can overcome intermediate node failures.

The rest of the paper is organized as follows. Section II presents the networking model overview and hierarchy. Section III presents the node addressing scheme and routing protocols. Section IV presents the simulation and analysis of results. Section V presents additional optimizations that can be used to further improve the performance of the network. The conclusions and future research are presented in the last section.

2   Networking Model Overview and Hierarchy

In this section, the architectural model of the sensor network is presented.

2.1   Node hierarchy

In the hierarchical model used, three types of nodes are defined:

- **Basic Sensor Nodes (BSN):** These are the most common nodes in the network. Their function is to perform the sensing function and communicate this information to the data relay nodes.
- **Data Relay Nodes (DRN):** These nodes serve as information collection nodes for the data gathered by the sensor nodes in their one-hop neighborhood. The distance

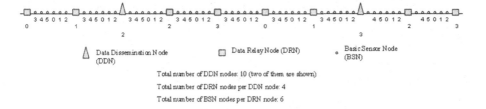

Fig. 1. Illustration of the addressing scheme used to assign DDN, DRN, and BSN address field values.

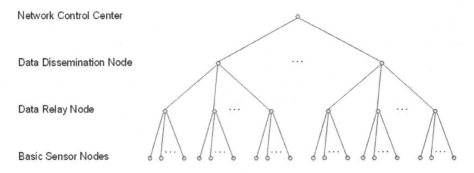

Fig. 2. A hierarchical representation of the linear structure sensor network, showing the parent/child relationship of the various types of nodes.

between these nodes is determined by the communication range of the networking MAC protocol used.

- **Data Discharge Nodes (DDN):** These nodes perform the function of discharging the collected data to the **Network Control Center (NCC)**. The technology used to communication the data from these nodes to the NCC center can vary. Satellite cellular technology can be used for example. This implies that each of the DDN nodes would have this communication capability.

The DDN nodes provide the network with increased reliability since the collected sensor data would not have to travel all the way along the length of the pipeline from the sensing source to the DRN center. This distance is usually very long and can be hundreds of kilometers. This would make it vulnerable to a large number of possible failures, unacceptable delay, higher probability of error, and security attacks. The DDN nodes allow the network to discharge its sensor data simultaneously in a parallel fashion. Additionally, the distance between the DDN nodes is important and affects the reliability of the network. A small distance between the DDN nodes would increase the equipment cost of the network, as well as deployment and maintenance costs. On the other hand a distance that is too large would decrease the reliability, security, and performance of the network. Figure 1 shows a graphic representation of the different types of nodes and their geographic layout. Figure 2 shows the hierarchical relationship between the various types of nodes in the sensor network. As shown in the figure, multiple BSN nodes transmit their data to one DRN node. In turn, several DRN nodes transmit their data to a DDN node. Finally, all DDN nodes transmit their data to the network control center.

3   Node Addressing Scheme and Routing Protocols

In order to facility routing, a multi-layer addressing scheme is used. The following section describes the address assignment process.

3.1   Multi-layer addressing

The logical address of each node consists of three fields. Hexadecimal or dotted decimal notation can be used for these fields. The order of the fields is: *DDN.DRN.BSN*.

- *DDN address field:* If this is a BSN or a DRN node, then this field holds the address of its parent DDN node. Otherwise, if this is a DDN node this holds its own address.
- *DRN address field:* If this is a BSN node, then this node holds the address of its parent DRN node. If this is a DRN node, then this field holds its own address. If this is a DDN node then this field is empty (i.e. holds a code representing the empty symbol, ϕ).
- *BSN address field:* If this is a BSN node, then this node holds its own address. If this is a DRN or DDN node then this field is empty.

A typical full address for a BSN node would be: 23.45.19. This means that its own BSN ID is 19, its parent DRN node ID is 45 and its parent DDN node ID is 23. A typical full address for a DRN node is: 23.45.ϕ. The empty symbol in the BSN field alone indicates that this is a DRN node. Finally, a typical full address for a DDN node is: 23.ϕ.ϕ. The two empty symbols in both the BSN and DRN fields indicate that this is a DDN node.

3.2   Address assignment

In this section, the process of assigning values to the different fields of the address of each node is described. Figure 1 shows an example linear alignment of DDN, DRN, and BSN nodes with the corresponding addresses for each node. The addresses in each field are assigned in the following manner:

- *DDN address field assignment:* The DDN nodes have a DDN address field starting at 0, 1, and so on up to (NUM_OF_DDN -1).
- *DRN address field assignment:* Each DRN node has as its parent the closest DDN node. This means that the DRN nodes belonging to a particular DDN node are located around it with the DDN node being at their center. The address fields of the DRN nodes on the left of the DDN node are assigned starting from 0, at the farthest left node, to (NUM_DRN_PER_DDN/2-1), where NUM_DRN_PER_DDN is the number of DRN nodes per DDN node. The address fields of the DRN nodes on the right start from (NUM_DRN_ PER_DDN/2) to (NUM_DRN_ PER_DDN - 1).

– *BSN address field assignment:* The BSN addressing field assignment is similar to that of the DRNs with the DRN node being that parent in this case. Each BSN node has as its parent the closest DRN node. This means that the BSN nodes belonging to a particular DRN node are located around it with the DRN node being at their center. The address fields of the BSN nodes on the left of the DRN node are assigned starting from 0, at the farthest left node, to (NUM_BSN_PER_DRN/2-1), where NUM_BSN_PER_DRN is the number of BSN nodes per DRN node. The address fields of the BSN nodes on the right start from (NUM_BSN_ PER_DRN/2) to (NUM_BSN_ PER_DRN - 1).

3.3   Communication between different types of nodes

– *Communication from BSN to DRN nodes:* As mentioned earlier each BSN node is within range of at least one DRN node. The BSN node will sign up with the closest DRN node. Subsequently, the BSN nodes transmit their information to the DRN node periodically. They also can be polled by the DRN node when the corresponding command is issued from the command center.
– *Communication from DRN to DDN Nodes:* Communication between the DRN and DDN nodes is done using a multi-hop routing algorithm which functions on top of a MAC protocol such as Zigbee. In this paper three different routing protocols for multihop communication among the DRN nodes are presented. These protocol are discussed later in this paper.
– *Information discharge at DDN nodes:* Collected data at the DDN nodes can be transmitted to the NCC center using different communication technologies. This implies that different DDN nodes would have different communication capabilities to transmit their collected information to the NCC center, depending on their location. For example nodes that are located within cities can send their information via available cellular GSM, or GPRS networks. On the other hand, nodes which are located in remote locations far from larger metropolitan areas might not be able to use standard cellular communications and would have to rely on the more expensive satellite cellular communication for transmission of their data. Another alternative would be to deploy WiMax or other long range wireless network access points at each 30 Km of the designated area along the pipeline.

3.4   The routing algorithms at the source and intermediate DRN nodes

As mentioned earlier each BSN node is within range of at least one DRN node. The BSN node will sign up with the closest DRN node. Subsequently, the BSN nodes transmit their information to the DRN node periodically. They also can be polled by the DRN node when the corresponding command is issued from the command center.

As mentioned earlier, when the DRN node is ready to send the data collected from its child-BSN nodes, it uses a multi-hop approach through its neighbor DRN node to reach its parent DDN node. The multihop algorithm uses the addressing scheme presented earlier in order to route the DRN packet correctly. Each DRN node keeps track of its connectivity to its neighbors through the periodic broadcast of hello messages among the DRN nodes. In order to increase network reliability, if the connection with the next hop is not available then the DRN node can execute one of three algorithms to overcome this problem.

Jump Always Algorithm (JA):

In order to still be able to transmit its DRN data successfully despite the lack of connectivity to its immediate neighbor, the DRN node can increase its transmission power and double its range in order to reach the DRN node that follows the current one. If multiple consecutive links are lost, then the DRN node can increase its transmission range appropriately in order to bypass the broken links. This process can happen until the transmission power is maximal. If even with maximal transmission power the broken links cannot be bypassed, then the message is dropped. In the protocol, this maximal DRN transmission power is represented by a network variable named MAX_JUMP_FACTOR which holds the maximum number of broken links or "disabled nodes" that a DRN transmission can bypass.

Redirect Always Algorithm (RA):

In this variation of the routing protocol, the DRN source node sends its DRN data message to its parent DDN node. While the message is being forwarded through the intermediate DRN nodes, if it reaches a broken link then the following steps are taken. The DRN node determines if this data message has already been redirected. This is determined by checking the *redirected* flag that resides in the message. If the redirected flag is already set then the message is dropped and a negative acknowledgement is be sent back to the source. Otherwise, the source can be informed of the redirection process by sending a short redirection message with the redirected message ID back to the source. The source will then re-send the data message in the opposite direction and update its database with the fact that this direction to reach the DDN node is not functional. Furthermore, in order to make the protocol more efficient the entire data message is not sent back to the source since the source already has a copy of the data message. Only a short redirection message with the redirected message ID is sufficient to be sent back to the source. Additionally, the redirection message also informs the other nodes on that side that there is a "dead end" in this direction and data needs to be transmitted in the other direction even if the number of hops to reach the other nearest DRN node is larger. In that case, each DRN node that receives this message will check the *redirected* flag, and if it is set, then it will continue to forward the message in the same direction. However, in order to prevent looping, if another broken link is encountered in the opposite direction the redirected message cannot be redirected again. In that case, the message is simply dropped.

Smart Redirect or Jump Algorithm (SRJ):

This algorithm is a combination of the first two algorithms JA and RA. We define as *sibling DRN nodes* to a particular DRN node x, the DRN nodes that have the same parent DDN node as x. We also define as *secondary sibling DRN nodes* to x the DRN nodes that have as parent DDN node the secondary parent DDN node (i.e. the DDN node that is on the opposite side of the parent DDN node with respect to x) of x. In this algorithm, each node contains information about the operational status of its sibling and secondary sibling DRN nodes. Consequently, before dispatching the message, it calculates the total necessary energy it needs to reach its parent DDN node E_x^p and the total energy it needs to reach its secondary parent E_x^{sp}. It then dispatches the message in the direction which takes the lower total energy to reach either the parent DDN or the secondary parent DDN. Specifically, if $E_x^p \leq E_x^{sp}$ then the message is sent towards the parent DDN node. Otherwise, the message is sent towards the secondary parent DDN node. This algorithm relies on the information in the node to reduce the total energy consumed by the network for the transmission of the message. This information about failure status of DRN nodes is cached by the DRN nodes from participations in previous packet transmissions. More research is being conducted for the most efficient means of gathering such information by the DRN nodes.

Table 1. Simulation Parameters

Parameter	Value
Total Number of DDN Nodes	5
Total Number of DRN Nodes Per DDN Node	100
Total Number of BSN Nodes Per DRN Node	6
DRN Transmission Rate	2 Mb/s
Periodic Sensing Interval	10 s
DRN Data Packet Size	512 bytes
MAX_JUMP_FACTOR	3

4   Simulation

Simulation experiments were performed in order to verify the operation, and evaluate the performance of the proposed framework and networking protocol. As indicated in Table 1, the number of DDN nodes used in the simulation is 5, and the number of BSN nodes per DRN node is 6. The number of DRN nodes per DDN node was varied between 100, 120, and 140. The results are presented in figure 3. For the JA and SRJ algorithms the MAX_JUMP_FACTOR is set to 3. All nodes are assigned their hierarchical addresses according to the addressing scheme that was discussed earlier. In the simulation, the BSN nodes send their sensed data to the their parent DRN node in a periodic manner. Then, the DRN nodes use the networking protocol to route this information to their parent DRN node. In order to verify and test the JA, RA, and SRJ routing protocols and their ability to route the generated packets correctly to the DDN nodes using intermediate DRN nodes, a number of DRN failures were generated using the Poisson arrival distribution with a certain average arrival rate. The average arrival rate of the DRN failures was varied in order to verify the addressing scheme and evaluate the capability of the routing protocol to overcome intermediate DDN node failures. As DRN nodes fail, routing of the DRN packets to either the parent DDN node or the alternative one in the opposite direction is done. When a DRN node fails, the three routing protocols react differently to overcome the failures as specified earlier in the paper. In this simulation, we are focusing on testing the correctness of operation of the protocols and assessing their performance with respect to each other. In figure 3, the number of DRN

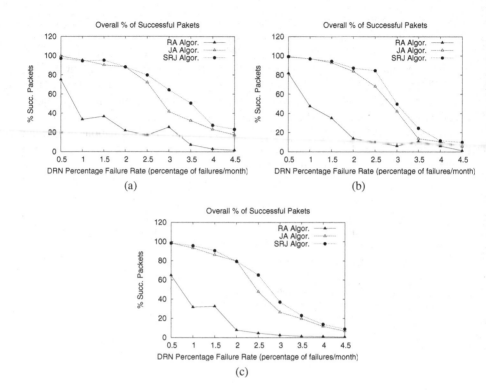

Fig. 3. Simulation results. (a) NUM_DRN_PER_DDN=100. (b) NUM_DRN_PER_DDN=120. (c) NUM_DRN_PER_DDN=140.

nodes per DDN node was varied in order to study the impact of increasing the number of DRN nodes per DDN node on network performance. The percentage of successfully transmitted packets was measured as the DRN percentage failure rate (percentage of DRN failures per month) was varied. As can be seen in all three parts of figure 3, the percentage of successfully transmitted packets decreases as the percentage of DRN failures increases. Also, it can be clearly seen that the SRJ algorithm provides the best performance followed by the JA algorithm and the RA algorithm respectively. This is expected since the RA algorithm does not try to jump over a failed DRN node, and only tries redirecting the packet once. If it encounters another failed DRN node in the opposite direction then the packet is dropped. The performance of the JA algorithm is better than that of the RA algorithm. This is also expected since the JA algorithm allows a DRN transmission to overcome failed nodes by jumping over them. However, if more than maximum number consecutive failed DRN nodes is encountered, then the packet is dropped without trying to go in the opposite direction, which might ensure successful transmission of the packet. The SRJ algorithm offers the best performance since it considers both directions and dispatches the packet only in the direction with the smallest required energy. In addition to providing more alternatives for overcoming failed DRN nodes, the SRJ algorithm also ensures a smaller number of DRN failures due to battery depletion which increases network lifetime and improves its performances.

Additionally, the results show that as the number of DRN nodes per DDN node increases from 100 to 120, to 140, the percentage of successfully transmitted packets decreases for all three algorithms. For example, for the SRJ algorithm case, with a percentage failure rate of 3 percent failures per month, the percentage of successfully transmitted packets decreases from 64.35 for DRN_PER_DDN = 100, to 49.72 for DRN_PER_DDN = 120, to 37.04 for DRN_PER_DDN = 140. This decrease in performance as the number of DRN nodes per DDN node increases is expected due to the linear structure of the network. With the increased number of DRN nodes that a packet has to use to reach the DDN node, the probability of encountering a more than maximum number of consecutive failed DRN nodes which prevents it from going further increases. Therefore, when designing such a network, the number of DRN nodes per DDN node must not be too large in order to ensure good network performance.

5   Additional Optimizations

This section presents some of the issues, observations, problems, possible solutions, and optimizations that are being considered for current and future research this area.

5.1   DRN types of failures

Two types of failures of DRN nodes can be identified depending on the cause of the failure:

– Normal-life DRN failures: These failures are due to the normal battery depletion of the DRN nodes.
– *Sub-normal-life DRN failures:* These failures are due to the expiration of a DRN node due to factors other than normal battery depletion from energy consumption. Such failures can be caused by physical damage, environmental damage, a manufacturing defect, a hardware or software problem, and so on. Such failures can happen at any time and to any DRN node regardless of its position with respect to the other nodes in the network. These failures can cause a black hole effect when using certain routing protocols. These effects will be discussed in a later section.

5.2   Proportional depletions: the suspension bridge effect

One important observation that is noted is that the energy consumption is higher in the DRN nodes that are closer to the DDNs. Specifically, the total energy consumption in the DRN nodes is inversely proportional to the number of hops of that node to the DDN nodes. This is due to the fact that as a node is closer to a DDN node, it will be a part of a proportionally higher number of paths from farther nodes that are trying to reach the DDN node. In other words, more farther nodes will use it as an intermediate node send their messages to the DDN node. This means that assuming that all DRN

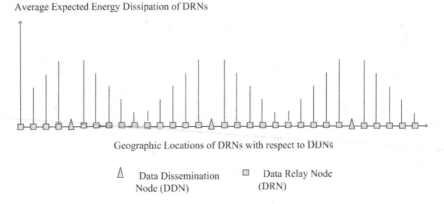

Fig. 4. Illustration of the suspension bridge effect: the average expected energy dissipation of DRN nodes according to their respective distance from DDN nodes.

node failures are due to normal-life failures, the DRN nodes that are within one hop of the DDN nodes will fail first, followed by the DRN node within 2 hops of the DDN nodes, then by the nodes within 3 hops and so on. If one is to plot the average expected energy dissipation of DRN nodes (on the y-axis) versus distance (on the x-axis) we get a suspension bridge-like figure where average expected energy dissipation of the one-hop DRN nodes (one hop from the DDN nodes) is the highest, followed by the 2-hop DRN nodes, and so on. figure 4 Shows an illustration of the average expected energy dissipation requirements of DRN nodes according to their respective distance from the DDN nodes. The figure shows that the closer a DRN node is to the DDN nodes the higher its average expected energy dissipation requirements.

5.3   Possible remedies to the suspension bridge effect

Two possible solutions can be used to remedy the rapid expiration of the DRN nodes closest to the DDN nodes.

Variable distance between DRN nodes: One solution would be to exponentially decrease the distance between DRN nodes as they get closer to the DDN nodes. This decrease in distance would require them to spend less energy to hop to the next node on the way to the DRN node. This will compensate for the higher number of transmissions that this node must do being a part of more paths that go through it. The change in the density of the DRN nodes can be done in such a way that the total energy consumption of the DRN nodes is the same regardless of their proximate position from the DDN nodes.

Variable initial energy capacity of DRN nodes: Another possible solution for this problem is to simply equip the nodes closer to the DDN with higher initial energy assuming. This solution is only possible if such feature or option is available with the type of technology and product that is used to implement the DRN nodes.

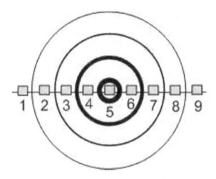

Fig. 5. An illustration of the black hole effect.

5.4 Depletions around a failed node: the black hole effect

Another type of effect can happen when a DRN node has a sub-normal-life failure. Due to this type of failure's unpredictable causes, it can happen to any DRN node at any given location with respect the DDN nodes and at any time. If the routing protocol that is used overcomes such a failure by jumping over the failed DRN node then this requires a higher energy consumption for transmission from the two surrounding DRN nodes. This means that the expected battery lifetime of these nodes is shortened. When these two nodes fail due to their decreased battery lifetime, the live DRN nodes next to them must now multiply their transmission power in order to overcome multiple adjacent DRN nodes that failed, the original one and the one next to it. In turn this will increase their energy consumption and cause them to fail when their batteries are depleted. The DRN nodes that are next to the three failed DRN nodes will now have to spend even more energy to jump over them, and so on. We name this process the *black hole effect*. The initialled failed DRN node is compared to a black hole that causes other adjacent DRN nodes to fail, which subsequently cause nodes adjacent to them to fail and widening the diameter of the failed DRN nodes. This is comparable to a widening black hole of failing DRN nodes with the initial failed DRN node at is center. This effect is illustrated in figure 5. In the figure, DRN node 5 fails first, and starts the process which causes the surrounding nodes to fail in sequence according to their proximity to the initial failed node. In this case, the failure of node 5 is followed consecutively by the failures of nodes 4/6, 3/7, and 2/8. This process continues until the number of failed adjacent DRN nodes is so high that it cannot be overcome by the maximum transmission range of DRN nodes. This results in partitioning of the linear network at this location.

6 Conclusions and Future Research

In this paper, an addressing scheme and routing protocol for linear structure wireless sensor networks was presented. This architecture and routing protocol are designed to meet the objectives of efficiency, cost-effectiveness, and reliability. The routing protocol is used to relay sensor information from the field nodes to a designated control

center. The protocol has the features of increased reliability by overcoming intermediate node failures, maximizing individual node battery life as well as extending network lifetime with minimal maintenance requirements. Simulation experiments were
conducted to test and evaluate the efficiency of the network protocol and underlying addressing scheme. Future work involves providing more detailed design and analysis of
the various aspects of the model, as well as further optimization of the routing protocol
and strategy. Security considerations will also be addressed and incorporated into the
design. In addition, more extensive simulation experiments will be conducted to evaluate the performance of the proposed model and its associated protocols under various
network conditions.

References

1. A. Carrillo, E. Gonzalez, A. Rosas, and A. Marquez. New distributed optical sensor for
   detection and localization of liquid leaks. *Pat I. Experimental Studies, Sens, Actuators*,
   A(99):229–235, 2002.
2. E. Fernandez, I. Jawhar, M. Petrie, and M. VanHilst. *Security of Wireless and Portable
   Device Networks: An Overview.*
3. I. Gerasimov and R. Simon. Performance analysis for ad hoc QoS routing protocols. *Mobility
   and Wireless Access Workshop, MobiWac 2002. International*, pages 87–94, 2002.
4. S. Hadim, J. Al-Jaroodi, and N. Mohamed. Trends in middleware for mobile ad hoc networks. *The Journal of Communications*, 1(4):11–21, July 2006.
5. Y. Hwang and P. Varshney. An adaptive QoS routing protocol with dispersity for ad-hoc
   networks. *System Sciences, 2003. Proceedings of the 36th Annual Hawaii International
   Conference on*, pages 302–311, January 2003.
6. I. Jawhar, N. Mohamed, and K. Shuaib. A framework for pipeline infrastructure monitoring
   using wireless sensor networks. *The Sixth Annual Wireless Telecommunications Symposium
   (WTS 2007), IEEE Communication Society/ACM Sigmobile, Pomona, California, U.S.A.*,
   April 2007.
7. I. Jawhar and J. Wu. Qos support in tdma-based mobile ad hoc networks. *The Journal of
   Computer Science and Technology (JCST)*, 20(6):797–910, November 2005.
8. I. Jawhar and J. Wu. Race-free resource allocation for QoS support in wireless networks. *Ad
   Hoc and Sensor Wireless Networks: An International Journal*, 1(3):179–206, May 2005.
9. I. Jawhar and J. Wu. Resource allocation in wireless networks using directional antennas.
   *The Fourth IEEE International Conference on Pervasive Computing and Communications
   (PerCom-06), Pisa, Italy. Publisher IEEE Computer Society*, pages 318–327, March 2006.
10. W.-H. Liao, Y.-C. Tseng, and K.-P. Shih. A TDMA-based bandwidth reservation protocol
    for QoS routing in a wireless mobile ad hoc network. *Communications, ICC 2002. IEEE
    International Conference on*, 5:3186–3190, 2002.
11. S. Nelakuditi, Z.-L. Zhang, R. P. Tsang, and D.H.C. Du. Adaptive proportional routing: a
    localized QoS routing approach. *Networking, IEEE/ACM Transactions on*, 10(6):790–804,
    December 2002.

Non-Custodial Multicast over the DTN-Prophet Protocol

José Santiago, Augusto Casaca and Paulo Rogério Pereira

INESC-ID/IST
R. Alves Redol, 9, 1000-029 Lisboa, Portugal
e-mails: {jose.santiago, augusto.casaca, paulo.pereira}@inesc-id.pt

Abstract. Networks with frequent and long duration partitions prevent common Internet protocols from working successfully. For protocols to work properly in these Delay/Disruption Tolerant Networks (DTNs), a new protocol layer was proposed that acts on top of the transport layer for the end-to-end exchange of messages (called bundles) taking advantage of scheduled, predicted, opportunistic or permanent connectivity. In this paper, we propose and evaluate a multicast extension to the DTN's unicast PROPHET protocol. A multicast protocol is useful to reduce the number of copies of packets when they are sent to multiple destinations. We show by simulation that by using just one byte for transferring mobility information between nodes, a good clue about the region where mobile nodes are is given, which can be used by the multicast protocol to decide where to forward messages. Additionally, we show that if the number of contacts between nodes is above a minimum threshold, a pseudo multicast tree will exist, multicast works efficiently and message replications are minimized.

Key words: Delay/Disruption Tolerant Networks, Connection Disruption, Multicast.

1 Introduction

Delay Tolerant Networks (DTNs) are networks that may experience frequent and long duration partitions. This occurs in situations in which no stable infrastructure exists that can guarantee permanent link connectivity. Examples of these situations are: military communications in the battlefield, deep space communications and rescue actions in catastrophe hit areas.

The Internet protocols are not useful for DTNs because link disruptions are not properly handled, causing protocols to timeout and abort. The DTN Research Group (DTNRG), which was chartered as part of the Internet Research Task Force (IRTF), has proposed an architecture [1] and a communication protocol [2] (the Bundle Protocol) for DTNs.

In DTNs, a message-oriented overlay layer called "Bundle Layer" is added. The Bundle Layer exists above the transport (or other) layers of the networks it interconnects. Application data units are transformed by the Bundle Layer into one or more protocol data units called "bundles", which are forwarded by DTN nodes according to the Bundle Protocol. To help routing and scheduling decisions, bundles

Please use the following format when citing this chapter:

Santiago, J., Casaca, A. and Pereira, P.R., 2008, in IFIP International Federation for Information Processing, Volume 264; Wireless Sensor and Actor Networks II; Ali Miri; (Boston: Springer), pp. 197–208.

contain an originating timestamp, useful life indicator, a class of service designator and a length indicator. The Bundle Protocol includes a hop-by-hop transfer of reliable delivery responsibility, called bundle custody transfer, and an optional end-to-end acknowledgement. Persistent storage may be used in DTN nodes to help combat network interruption.

The Bundle Protocol does not include a bundle routing protocol nor mechanisms for populating the routing or forwarding information bases of DTN nodes. These functions are left for protocol extensions or for other protocols. One of the important routing protocols in DTNs is Epidemic Routing [3], which works by flooding the network with the messages. Although it provides the optimal solution for DTNs as regards the delivery ratio and latency, it is very wasteful of resources.

The PROPHET (Probabilistic ROuting Protocol using History of Encounters and Transitivity) protocol [4] is a routing protocol for unicast communication in DTNs. The PROPHET Protocol (DTN-PP) uses the history of encounters between nodes and transitivity to estimate the probability of nodes meeting and exploits the mobility of some nodes to bring messages closer to their destination. The DTN-PP is an alternative to Epidemic Routing with lower demands on buffer space and bandwidth, with equal or better performance in cases where those resources are limited and without loss of generality for scenarios where it is applicable.

Multicast communications are used when data is to be sent efficiently and simultaneously to a group of destinations, creating copies of the data not in the source, but only as required by the paths to the destinations. Extensions to the Bundle Protocol for supporting multicast are defined in [5]. No multicast routing strategies are yet defined, just the basic mechanisms for supporting multicast. An overview of multicast models for DTNs is presented in [6], but no specific protocol is proposed.

In [7], we proposed the Multicast over DTN-Prophet Protocol (MoDTN-PP) as an extension to DTN-PP for non-custodial multicast. Non-custodial means that the protocol will do its best effort to deliver messages. No node assumes custody for bundles being transmitted to the destination, so no special actions will be done to assure success. The probabilistic model from DTN-PP for node contacts was kept in MoDTN-PP. A pseudo-multicast tree mechanism was added to manage multicast groups.

In this paper, we describe how indications of the location and direction of the moving nodes are introduced in MoDTN-PP to help forming a pseudo multicast tree. We show that DTNs can benefit with non-custodial multicast communications when certain requirements are respected. We demonstrate that if there are a minimum number of contacts between nodes, multicast works efficiently, minimizing the number of message replications done in the network.

Section 2 of the paper summarizes the main aspects of probabilistic routing in DTN-PP. Section 3 provides an overview of the DTN-PP extensions for multicast. Sections 4 and 5 are dedicated to the simulation goals and results. Finally, section 6 presents conclusions and further work topics.

2   The DTN Prophet Protocol Probabilistic Model

DTN-PP uses the concept of probabilistic routing instead of epidemic routing, exploring the predictability of opportunistic contacts for message dissemination. In DTN-PP, the calculation of the delivery predictability is done in three parts as shown in the equations below. $P(x,y)$ is the delivery or contact probability from node x to node y, which is used as a routing metric.

$$P_{(a,b)} = P_{(a,b)_{old}} + (1 - P_{(a,b)_{old}}) \times P_{init}; \qquad\qquad 0 \leq P_{init} \leq 1. \qquad (1)$$

$$P_{(a,b)} = P_{(a,b)_{old}} \times \gamma^k; \qquad\qquad 0 \leq \gamma < 1. \qquad (2)$$

$$P_{(a,c)} = P_{(a,c)_{old}} + (1 - P_{(a,c)_{old}}) \times P_{(a,b)} \times P_{(b,c)} \times \beta; \qquad 0 \leq \beta \leq 1. \qquad (3)$$

Equation (1) is for probability update whenever two nodes meet: it will increase $P(x,y)$ every time they meet, since nodes that contact frequently have a higher probability to exchange messages. $P(x,y)_{old}$ is the last previous calculation in node x. P_{init} is an initialization constant that controls the rate of probability increase.

Equation (2) lowers the $P(x,y)$ probability as time passes, being used whenever the tables are updated. If contacts are rare, the probability should be reduced to reflect this. γ is an aging constant. k is the number of time units that have elapsed since the last time the metric was updated.

Equation (3) expresses transitivity, as messages can go from "a" to "c" directly or via "b". If "b" frequently meets "c", then it is probably a good node for forwarding messages to "c". β is a scaling constant that models the impact the transitivity should have on the delivery predictability.

This probabilistic model was kept in MoDTN-PP.

3   Multicast over DTN-Prophet Protocol

MoDTN-PP adds multicast message delivery capabilities to DTN-PP. MoDTN-PP introduces the use of mobility information, which is an innovative point compared with DTN-PP. This means that MoDTN-PP can use the information of the node position and of its moving direction in addition to the estimate of the probability of contact between nodes from DTN-PP.

3.1   Mobility Information Mechanism

Contacting nodes need to exchange the following control information:
- Contact probability of nodes which move in the same area;
- Their own geographical position and the geographical position of the nodes they contacted;
- Their moving direction and the moving direction of the nodes they contacted.

This additional information can be carried in just one byte. The convention used is as follows:

– The two most significant bits carry the direction information;
– The three following bits carry the x coordinate;
– The last three bits carry the y coordinate.

The moving direction of a node identifies the approximate direction the node is heading as the quadrant to which the node is moving to (identified with a number between 0 and 3). If the node is stopped or moving in the border between quadrants, then any quadrant may be reported. Of course, the moving direction can change, but this is not known beforehand. This relevant information is complemented with the approximate global position coordinates (referred to the total area), at the moment in which the node contact happens.

With this method, the mobile's position and direction passed to a peer are not very accurate. However, for most situations it is enough to introduce efficiency in the system. Indeed, nodes can be always moving and connections are difficult. Their actual position cannot be delivered instantaneously to the entire network. When this information is propagated, it may become obsolete very fast, so it is not very important to be accurate. As a matter of fact, a system will be more cost-effective if it consumes fewer resources and just gives good clues about the region where a mobile node shall be found. In addition, unless mobile nodes have GPS (Global Positioning System), they may only know the region they are in by identifying some wireless access point in the neighbourhood or by getting their position from adjacent nodes.

3.2   Information Exchanged During Nodes Contact

When nodes establish contact, they exchange two types of information: routing and data.

All the mechanisms related with unicast routing from DTN-PP were kept in the multicast extension. Nodes exchange a Routing Information Base Dictionary (Dict) that includes the address list used to make routing decisions for all known destinations. The mobility information in MoDTN-PP was added to this dictionary. Nodes also exchange their Routing Information Base (RIB), where contact probabilities are stored.

After the exchange of Dict and RIB messages, bundles with data messages start to be exchanged. With the introduction of multicast traffic, multicast and unicast bundles are stored in separate queues. As a consequence of this option, multicast bundles are offered after the unicast bundles and are requested in the same order. With this method, unicast traffic has some kind of priority as compared with multicast traffic.

Some small modifications were made to the DTN-PP state machines to stop sending periodic messages when there is no more useful information to exchange, saving energy.

3.3   Pseudo Multicast Tree

In traditional multicast, a well-defined tree is created whenever a group interested in the communication is active. This tree connects all members of the multicast group to a point in the network. This point can be the source or another point called rendezvous point (RP). In the first case, the tree is called a source-based tree and in the second case it is known as a shared tree.

For DTNs, such trees cannot be created permanently. Connections between nodes may have a very short duration, so an all-way active path from the source (or RP) to one or more destinations through a considerable number of nodes is not expectable. As a consequence, in DTNs, the concept of a multicast tree is broader. This pseudo multicast tree will be composed by one source (or RP) and a set of destinations with a set of intermediate nodes (mules) between them. Mules get data packets from sources - or RPs - or other mules and forward them to destinations points.

When a node needs to join a multicast group, it uses the best of its neighbours to contact the group. The choice is made with a heuristic based on the conditions of neighbours as shown in Table 1. Every condition is evaluated based on the available information at the moment. The weight of every condition that is true is added. The neighbour that adds up more is selected. The weights were chosen to value connections as direct as possible to the source. This is the reason the largest weight (30) is for a direct connection to the multicast source. The second largest (14) is for a neighbour already contributing to the multicast tree operation. The third weight (10) is for a neighbour that can contact the multicast source with high probability. The forth weight (6) is to distinguish from the case where this probability is zero. The remaining conditions are for giving preference to good contacts with neighbours, to neighbours near the source, to neighbours moving near the source or to neighbours that can contact any member of the multicast group. The study of other parameters or heuristics was left for further work.

Table 1. Criteria for selecting a neighbour for inclusion on the multicast tree

Weight	Condition of neighbour
30	is the source of the group
14	already serves as a mule
10	has/can contact the source with $P_{(x,y)}>=0.7$
6	has/can contact the source with $0<P_{(x,y)}<0.7$
4	is contacted with $P_{(x,y)}>0.7$
3	is near the source
2	moves towards the source, or neighbour and source go in the same direction
1	can contact the targeted multicast group

By adding some location information to the routing information exchanged, the nodes have a reasonable probability of knowing where the multicast source is. This helps to reduce the size of the pseudo multicast tree and optimizes the paths followed by messages.

When the connection to the group fails, the group member searches in his pool of neighbours for another node that can provide this service.

3.4   Group Membership

The group membership mechanisms of MoDTN-PP were previously described in [7]. A few control messages for dynamically managing multicast groups were added to the Bundle Protocol [2]. A JOIN message is used to join a multicast group. It will be propagated until reaching a group member, making the nodes in this path candidates to become mules. A LEAVE message may be sent to explicitly leave the group. Several timeouts exist to control the mechanisms and free resources when communication is no longer possible. Also, a new field for multicast mode usage was proposed in [8] to be added to the Bundle Protocol messages. This field carries the previous hop identity, which permits a node to determine the shortest path to the source. In this way, loops can be prevented. This field was also added to the MoDTN-PP implementation.

4   Simulation Conditions and Goals

In order to evaluate MoDTN-PP, we had to recreate the conditions in which DTN-PP was tested. It is important to test MoDTN-PP in the same conditions in which DTN-PP proved to work well. In this way, it can both be observed if its performance was affected and if unicast and multicast modes perform satisfactorily.

4.1   Test Scenario Definition

Two scenarios used in the DTN-PP evaluation [9] were reproduced. The first scenario was generated using the Random Waypoint Mobility Model [10], which is a model commonly used by the scientific community, and the second scenario is based on the Community Model [9], which was originally created for DTN-PP tests. For the sake of the evaluation, the main parameters of the DTN-PP mobility models were preserved. This means that, for both cases, hereafter referred as Scenario_1 and Scenario_2, respectively, the parameters used and shown in Table 2 were strongly based on those defined in [9].

Table 2. Specifications of the scenarios

Parameters	Scenario_1	Scenario_2
Area	1500 m × 300 m	3600 m × 1500 m
Total number of nodes	50	50
Speed	0 – 20 m/s	10 – 30 m/s
Pause time	5 – 13 s	5 – 13 s
Warm up period	500 s	500 s
Message generation period	1980 s	3000 s
Simulation duration	4500 s	11500 s

Parameters in Table 2 have literal meanings. "Area" represents the geographical area in which mobile nodes move; "Speed" indicates the limits between which the mobile speed varies and "Pause time" is the time range during which the mobile stops when it arrives at a new position. "Warm up period" represents the moment at which messages begin to be generated after the start of the simulation (the warm up period is needed to allow the DTN-PP delivery probabilities to have initial values). The message generation stops when the "Message generation period" is elapsed.

Besides the differences between both scenarios shown in Table 2, there are other differences related to the mobility model used in each of the scenarios. In Scenario_1 all nodes move randomly. The movement of each node starts by randomly selecting a new direction and position. As the node arrives at that position, it stops for a period of time randomly selected between 5 and 13 seconds. After that, it repeats the procedure to move to another place. The Community model has a different philosophy. Although the node, after arriving at a certain position, also stops for a period of time between 5 and 13 seconds, in this case the geographic area of Scenario_2 is divided into 12 sub-areas as in [9]. Each represents a community place. The last sub-area is the "Gathering place". In each of the 12 sub-areas there is a community static node that helps message exchange between mobile nodes. Sub-areas from 0 to 10 are homes for the nodes. The number of resident nodes may vary from 3 to 6 in each sub-area. Finally, node destinations are selected probabilistically as shown in Table 3. Selecting a destination means: if a node is at "Home", 0.8 represents the probability of the "Gathering place" to be its next destination and 0.2 is the probability for the node to go elsewhere; in the same way, the probability for the node to go "Home" when it is "Elsewhere" is 0.9, and only 0.1 to move to another sub-area.

Table 3. Destination selection probabilities

From/To	Home	Gathering place	Elsewhere
Home	-	0.8	0.2
Elsewhere	0.9	-	0.1

The reasons for creating a scenario such as the Community Model are given by its authors. In their own words, this kind of scenarios, where mobility can occur, involves human mobility in communities represented by villages and larger towns. Furthermore, towns themselves can represent the gathering place to and from which people resident in the surroundings move to work every day. Another rural example used for this gathering place model is a feeding ground where shepherd communities conduct animals with sensors attached to them. Indeed, projects like ZebraNet and other related to semi-nomadic Saami population of reindeer herders, in the north of Sweden, were inspired for seemingly realities.

Other differences between the two scenarios are related to traffic. In Scenario_1 a message is sent every second from any node belonging to a subset of forty-five, as five nodes do not contribute for data traffic. If the message must be delivered in unicast mode, the source is selected randomly and the destination will be any node belonging to the remaining forty-four nodes. However, if the message must be delivered in multicast mode, the message is only distributed to the other members of the group (in our case, 8 nodes). A ninth node acts as the multicast group source. Only one multicast group was active in the simulations. The group is identified by the association of two addresses: the source address and the group address. In Scenario_2, multicast messages have the same source and end-destinations as in Scenario_1. However, unicast is generated in a different way when compared with the first scenario. In Scenario_2, two randomly selected community static nodes send one message every ten seconds for static nodes located in other communities. Every time, five seconds later, two randomly selected mobile nodes send one message to randomly selected destinations. The number of multicast bundles was configured to be 54.7% of the number of unicast bundles in Scenario_1 and 53.8% in Scenario_2. These values ensure a significant proportion of multicast, but still keeping unicast traffic dominant.

Finally, the parameter values used to test MoDTN-PP were kept exactly as in the original protocol scenarios. These parameters are defined in the protocol to calculate contact probabilities between nodes as seen in section 2. The parameter values used are shown in Table 4.

Table 4. MoDTN-PP parameters

Parameter	Value
P_{init}	0.75
β	0.25
γ	0.98

4.2   Measuring Goals

The main goals of the evaluation are to verify under which conditions multicast in MoDTN-PP works and to assess its performance as compared to unicast.

Being wireless communication a subjacent goal, the evaluation criterion used is based on the wireless link range. Considering the link range used by many equipments operating in the wireless area, one hundred meters was selected as an average value: 60m and 160m limit the test interval. In this kind of simulation, the variation of the link range corresponds directly to the variation of the number of node contacts. This effect could be also achieved by varying the number of nodes in each simulation. In both cases, the number of opportunities for nodes to exchange messages is linearly related to the number of nodes or to their wireless link range.

The performance is measured by the bundle delivery ratio (received messages/registered messages) and the average bundle delay.

As a supplementary goal of the evaluation, it is also important to analyse how the multicast trees succeed, if they succeed, while the system works. Indeed, a starting point for this analysis is the fact that multicast trees cannot permanently exist in presence of multiple and frequent link disruptions. This means that it is not expected to have an end-to-end path that ties the multicast group source to all group members, through some intermediate nodes. This is a principle for DTN networks, and cannot be ignored while trying to use the multicast mode. However, another system principle is that there are a large enough number of nodes that move. Moreover, there are regularities in these moves that facilitate encounters among nodes. Some sub-group of these nodes will maintain more or less longer contact as they move in a similar direction. If these same nodes are members of a multicast group, they can form a multicast tree branch acting as a bridge that paves the way for multicast bundles to go from a passing group member to another that casually contacts the opposite side of the branch.

5   Simulation Results

The simulations were done on the ProphetSim simulator [11], whose implementation is based on OMNeT++, version 3.2p1, and uses the Mobility Framework, version 1.0a6.

For evaluation purposes, the two scenarios described in section 4 have been used. In the graphs below, each point represented is the average result of five simulations. The same simulation was used for generating the multicast and unicast data, so that performance can be compared. Fig. 1 shows the bundle delivery ratio as the ratio between received messages and sent messages. If the message is not delivered to all group members by the end of the simulation, it counts to the delivery ratio with a value equal to the ratio between the number of group members that received the message and the total number of group members.

Fig. 2 shows the average delay for bundles to be delivered to their destination. This delay is an average delay for all destinations to which the bundles were successfully delivered within the simulation time.

Fig. 3 shows the number of nodes grafted to the multicast tree in a randomly selected starting point (186s) and with a randomly selected period (800s). This number includes the members of the multicast group (source plus 8 destinations) as well as mule nodes that cooperate in the message transmission.

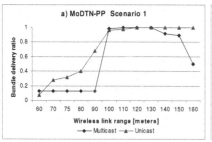 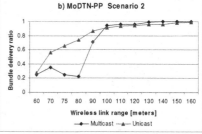

Fig. 1. Bundle delivery ratio for: a) Scenario_1; b) Scenario_2

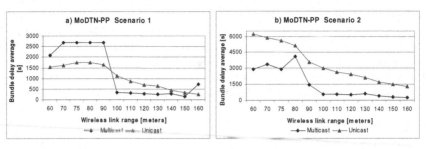

Fig. 2. Bundle delay average for: a) Scenario_1; b) Scenario_2

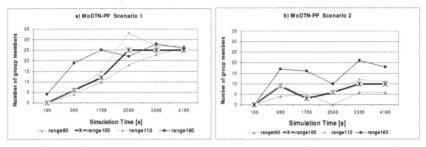

Fig. 3. Number of Multicast group members for: a) Scenario_1; b) Scenario_2

5.1   Wireless Link Range Analysis

The simulation results depend on the scenario characteristics. The results shown in Fig. 1 and Fig. 2 must be divided into three parts, according to the wireless link range, as the quality of a contact depends of signal propagation and contact duration:

1   *From 60m to 100m*: here, unicast performs badly and multicast performs worse. This is because contacts are infrequent and their quality is not good. This can be better observed in Fig. 1-a) for wireless ranges below 100 m, where multicast performance quickly drops to a very low value. In these conditions, multicast tree branches and multicast operations have almost no opportunity to work.

2   *From 100m to 130m*: both unicast and multicast perform well. Multicast presents better results for the bundle delay average. For this wireless link range, the number of contacts and their quality is good for both modes. It must be remembered that, generally speaking, the quality of these contacts are better than in the previous situation. In the simulation conditions, signal propagation is strictly related with link range; and the longer this range is, the greater is the duration of contacts between nodes.

3   *From 130m to 160m*: in this case, it could be expected that the increased number of contacts would guarantee good conditions. However, multicast performance declines. In fact, as in the simulator implementation unicast bundles are offered first than multicast bundles, the number of unicast bundles

is greater than multicast bundles and as the unicast bundles have no limits for replication in the simulator, the number of unicast bundle copies in the net becomes too large and multicast bundles starve. The quality of communication is, on average, the same as before, but it is not enough for multicast. Now the contact duration is not sufficient and the minimum number of contacting node pairs is not guaranteed for a significant number of mobiles. The large unicast bundle traffic even causes perturbations in the multicast tree. This can be observed in point 2586 s of Fig. 3-a), where the results for range 160 m are worse than for shorter ranges. This suggests that a scheduling mechanism that properly shares the bandwidth between unicast and multicast should be implemented. This was left for further work.

5.2   Results

The simulation results confirm that the use of multicast mode is advantageous. Moreover, Scenario_2 is positively thought to be more realistic than Scenario_1. It fits better to the human community activity, and it is here that multicast adds better quality of service. After achieving a sufficient number of contacts with quality, the bundle delivery ratio is the best in multicast.

The bundle delay average for multicast is even better than for unicast as can be observed in Fig. 2. This can be explained by the use of mobility information for the multicast routing, while unicast routing just uses contact probability. The mobility information permits the pseudo multicast tree to have more direct connections, optimizing the paths and reducing delay.

Figures 1 to 3 illustrate that multicast trees work in DTNs if the needed conditions exist: valuable contacts between nodes, with enough duration and communication possibility. Fig. 3 also shows that the effectiveness of the multicast tree is not proportional to the number of nodes grafted with the tree. Indeed, the multicast mode performs better in Scenario_2, despite the larger number of multicast members in Scenario_1. This happens because another factor affects the results. Indeed, the geographical position that crucial members occupy in the overall network is also important. A paradigmatic example is given by a node positioned in a strategic point, where many nodes pass towards multiple directions. This case represents privileged places where mules load, unload and exchange messages.

6   Conclusion

In DTNs, communication channels may fail repeatedly. They can do that for long periods of time. Considering such conditions, it might not seem realistic to try to use multicast trees to deliver messages in scenarios like these. We extended DTP-PP with a multicast mode, adding information with indications of node position and direction of movement to help forming pseudo multicast trees.

The work presented here proves that the multicast mode can be used if a minimum of contacts between nodes exists. The use of position and direction information proved to be a good routing metric, contributing to the existence of a pseudo-multicast tree, which results in shorter message transfer delays. There are situations in which multicast can even perform better than unicast, as is the case of Scenario_2. This is justified by the use of nodes located in strategic points that can be crucial for the performance of communication.

Multicast can improve DTNs efficiency, as it permits saving resources as the number of message replications is minimized in the network. Fewer replications mean that less processing capacities are needed, more memory and bandwidth are available and packets suffer less delay and loss. This adds a valuable contribution to the quality of service in DTNs as it does for other communication systems.

The decision of offering unicast bundles first than multicast bundles proved to be unfair to multicast traffic when there are frequent contacts between nodes. This suggests developing a scheduling mechanism to share bandwidth fairly between unicast and multicast as further work.

Some other further work topics are: testing other heuristics for building the pseudo-multicast tree; determining the best values for the protocol parameters according to the mobility model or automatically; identifying to which realistic mobility models multicast is better adapted; and having a validation method for information exchanged between nodes.

References

1. V. Cerf et al., "Delay Tolerant Network Architecture", IETF, RFC 4838, April 2007.
2. K. Scott and S. Burleigh, "Bundle Protocol Specification", IETF, RFC 5050, November 2007.
3. Vahdat and D. Becker, "Epidemic Routing for Partially Connected Ad Hoc Network", Duke University Technical Report Cs-200006, April 2000.
4. Lindgren and A. Doria, "Probabilistic Routing Protocol for Intermittently Connected Networks", IETF, draft-lindgren-dtnrg-prophet-02.txt, March 2006.
5. S. Symington, R. Durst, "Bundle Protocol Extensions to Support Multicasting", IETF, draft-irtf-dtnrg-bundle-multicast-00.txt, October 2005.
6. Zhao W., Ammar M., and E. Zegura, "Multicasting in Delay Tolerant Networks: Semantic Models and Routing Algorithms", SIGCOMM'05, Philadelphia, USA, August 2005.
7. José Santiago and Augusto Casaca, "Non-Custodial Multicast Operations on Intermittently Connected Networks", EuroFGI Workshop on IP and Traffic Control, pp.143-150, Lisbon, Portugal, December 2007.
8. S. Symington, R. Durst, K. Scott, "Delay-Tolerant Networking Previous Hop Extensions Header", IETF, draft-irtf-dtnrg-bundle-previous-hop-extension-header-00, April 2006.
9. Lindgren, A. Doria, and O. Schelén "Probabilistic Routing Protocol for Intermittently Connected Networks", Mobile Computing and Communication, Springer LNCS, Volume 3126, pp. 239-254, January 2004.
10. T. Camp, J. Boleng, and V. Davies, "A survey of mobility models for ad hoc network research," Wireless Communications & Mobile Computing (WCMC): Special issue on Mobile Ad Hoc Networking: Research, Trends and Applications, vol. 2, no. 5, pp. 483-502, 2002
11. Anders Lindgren Software Releases. http://www.sm.luth.se/~dugdale/index/software.shtml

Improving Mobile and Ad-hoc Networks performance using Group-Based Topologies

Jaime Lloret[1], Miguel Garcia[2] and Jesus Tomas[3]

Department of Communications
Polytechnic University of Valencia
Camino Vera s/n, 46022, Valencia (Spain)
[1]jlloret@dcom.upv.es, [2]migarpi@posgrado.upv.es, [3]jtomas@dcom.upv.es

Abstract. Many works related with mobile and ad-hoc networks routing protocols present new proposals with better or enhanced features, others just compare them or present an application environment, but this work tries to give another point of view. Why don't we see the network as a whole and split it intro groups to give better performance to the network regardless of the used routing protocol?. First, we will demonstrate, through simulations, that grouping nodes in a mobile and ad-hoc networks improves the whole network by diminishing the average network delay and also the routing traffic received by the nodes. Then, we will show which one of the actual fully standardized protocols (DSR [1], AODV [2] and OLSR [3]) gives better performance to the whole network when there are groups of nodes. This paper starts a new research line and urges the researchers to think on it and design group-based protocols.

Keywords: MANET, group-based topologies, network performance.

1 Introduction

The routing protocols in mobile and ad-hoc networks are divided into three types: proactive (which update the routing tables of all the nodes periodically), reactive (which maintain routing routes in their tables only when a node has to communicate with another node in the network) and hybrid (which are a combination of the other two types, taking the advantages of both types). There are many works in the literature that compare the performance of the routing protocols. The most compared protocols have been DSR and AODV. In references [4] and [5] we can see their comparison taking into account some parameters such as the packet delivery fraction, the average delay, the normalized routing load and the throughput consumed when the network load, the mobility and the network size vary. The work in reference [6] added the STAR protocol to the comparison and they measured the data delivery, the control overhead and the data latency. Reference [7] compared DSR and AODV with DSDV taking into account the average delay, the throughput and the control overhead with varied mobility. On the other hand, reference [8] compared DSR, AODV and TORA to analyze the control traffic sent, the data traffic received, the data traffic sent, the throughput, the retransmission attempts, the radio receiver throughput, the radio

Please use the following format when citing this chapter:

Lloret, J., Garcia, M. and Tomas, J., 2008, in IFIP International Federation for Information Processing, Volume 264; Wireless Sensor and Actor Networks II; Ali Miri; (Boston: Springer), pp. 209–220.

receiver utilization, the average power, the radio transmitter utilization, the radio transmitter throughput, routing traffic received, routing traffic sent, number of hops and route discovery time. The paper in reference [9] compares the number of packets sent and the traffic sent by DSDV, TORA, DSR and AODV protocols in networks of 50 mobile nodes. Other works compared 5 protocols such as the one presented in [10], where AODV, PAODV, CBRP, DSR and DSDV were compared taking into account the data packet throughput, the average data packet delay and the normalized packet overhead for various number of traffic sources.

Current IETF standardized protocols are AODV [1], DSR [2] and OLSR [3]. None of the works aforementioned have compared them from the group-based topology point of view. We are going to analyze and study their performance when there are group of nodes in their topology.

A cluster is made by a cluster head node, cluster gateways and cluster members. The cluster head node is the parent node of the cluster, which manages and checks the status of the links in the cluster, and routes the information to the right clusters. The rest of the nodes in a cluster are all leaf nodes. The size of the cluster is usually about 1 or 2 hops from the cluster head node. Cluster-based networks are a subset of the group-based networks, because every cluster could be considered as a group. But a group-based network is capable of having any type of topology inside the group, not only clusters. We will take care of group-based topologies in this paper.

The paper is structured as follows. Section 2 shows and analyzes the differences between DSR, AODV and OLSR protocols when regular and group-based topologies are used. The group-based topologies comparison is shown in section 3. Finally, section 4 gives our conclusions.

2   Group-based topology performance

2.1   Test Bench

This sub-section presents the test-bench used for all the evaluated protocols. The number of nodes and the coverage area of the network have been varied. Each protocol has been simulated in 4 scenarios: (1) With fixed nodes, (2) With mobile nodes and failures, (3) With grouped nodes and (4) With grouped mobile nodes and failures. Each scenario has been simulated for 100 and 250 nodes, to observe the system scalability. Instead of a standard structure we have chosen a random topology. Figure 1 shows the 100 nodes topology (in a 750x750 m^2 area) and Figure 2 shows the 250 nodes topology (in a 1 Km^2 area). It has been obtained using the version Modeler of OPNET simulator [11]. Both topologies have been created using different seeds. Arrows indicate that nodes are mobile and change their position constantly. The green lines from each node (blue circles) indicate the node mobility. We can see that the nodes are inside a blue box. This box shows a wireless area and it has been used to delimit the mobility area of the nodes. In that area, a node can move randomly during the simulation. The physical topology doesn't follow any known pattern. The obtained data don't depend on the initial topology of the nodes or on their movement pattern, because all of it has been fortuitous.

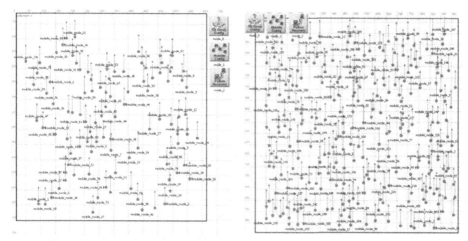

Fig 1. Topology with 100 nodes.   **Fig 2.** Topology with 250 nodes.

We have created 6 groups for the 100 nodes topology, covering approximately, a circular area with a 150 meter radius each group. There are 16 or 17 nodes approximately, in each group. The number of nodes in each group varies because of the node's random mobility. A node can change a group anytime. For the 250 nodes topology, we have created 12 groups, with 15 or 16 nodes per group approximately covering a circular area with a 150 meter radius each group.

The ad-hoc nodes of the topologies have a 40 MHz processor, a 512 KB memory card, a radio channel of 1 Mbps and their working frequency is 2.4 GHz. Their maximum coverage radius is 50 meters. This is a conservative value because most of the nodes in ad-hoc network have larger coverage radius, but we preferred to have lower transmitting power for the ad-hoc devices to enlarge their time of life.

We have forced node failures at t=200 sec., t=400 sec. and t=1200 sec. in each network, with a recovering process of 300 sec., to take measurements from the mobile nodes simulation when the physical topology changes.

The MANET traffic generated by OPNET has been used as the simulations' traffic load. We inject this traffic 100 seconds after the simulation starts. We have configured the traffic arrival with a Poisson distribution (with a mean time between arrivals of 30 seconds). The packet size follows an exponential distribution with a mean value of 1024 bits. The destination address of the injected traffic is random to obtain a simulation independent of the traffic direction. We have simulated the four scenarios for DSR, AODV and OLSR protocols. The results obtained are shown in the following sub-sections.

2.2   DSR, AODV and OLSR in group-based topologies

Figures 3 and 4 show the average delay of the DSR protocol in fixed and mobile topologies at the application layer. In figure 3 we observe that group-based topologies have an average delay close to 0.005 seconds regardless of the number of nodes in the network. In the regular network the delay has a value of 0.02 seconds for 100-nodes topology and of 0.03 seconds for the 250-nodes topology when the network converges. In the case of the 100-nodes topology there is an improvement of 75% and it is better in the 250-nodes topology (an 83% of improvement). The topologies with mobility and errors (figure 4) shows that the average delays at the application layer are higher in the group-based topologies till the network converges. Although group-based topologies present worse behaviour till 1300 seconds, when the network is stabilized, group-based topologies have an improvement around 5%.

Then, we have compared the routing traffic received in the DSR protocol (figures 5 and 6). Figure 5 shows that the traffic is quite stable due to the characteristics of the network. It is due to it is a fixed network without errors and failures. The traffic received in the 250-node topology is around 500 Kbits/s, but when we group the nodes this traffic decreases until 200 Kbits/s (a 60% of improvement). The value obtained in a 100-node topology (250 Kbits/s), is also improved when we group the nodes (100 Kbits/s), therefore there is a 60% of improvement. In figure 6 we observe a similar behaviour. In this case we conclude that when there are errors and failures in the 250-nodes topology the traffic fluctuates and is less stable (we can observe it in the intervals from 600 to 800 seconds and around 1200 seconds). We also observe that the instability is much lower in group-based topologies. 100-nodes topology has a mean value around 175 Kbits/s, while 100-nodes group-based topology has a mean value around 95 Kbits/s, so there is an improvement of 46%. On the other hand, 250-nodes topology has a mean value around 400 Kbits/s, while 250-nodes group-based topology has a mean value around 180 Kbits/s, so there is an improvement of 55%.

The average delay at the application layer in the AODV protocol can be seen in figures 7 and 8. Both topologies, 100-nodes and 250-nodes, give an average delay higher than 0.5 seconds when the network converges, but there are some peaks higher than 2.5 seconds. On the other hand, group-based topologies have a similar delay which is around 0.15 seconds. Group-based topologies improve the delay at the application layer in 70%. When the topology with mobile nodes is used, the simulation shown in figure 8 is obtained. In case of 250 nodes, there is a delay of 1 second when the network has converged. The case of 100 nodes gives an average delay of 0.75 seconds approximately. When there are group-based topologies, the delay decreases to 0.25 seconds in both cases. There is an improvement of 75% for the 250-nodes topology and 67% for the 100-nodes topology.

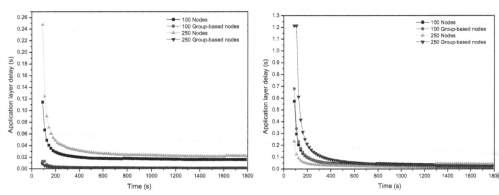

Fig. 3. DSR average delay at the application layer in fixed topologies.   **Fig. 4.** DSR average delay at the application layer i. mobile topologies.

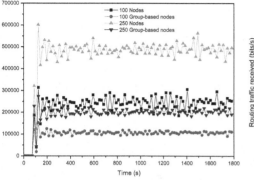

Fig. 5. DSR routing traffic received in fixed topologies.

Fig. 6. DSR routing traffic received in mobile topologies.

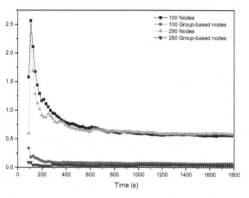
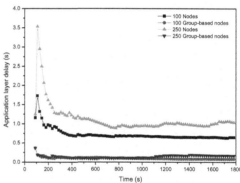

Fig. 7. AODV average delay at the application layer in fixed topologies.

Fig. 8. AODV average delay at the application layer in mobile topologies.

The routing traffic received for the AODV in each simulated topology can be seen in figures 9 and 10. We observe that the routing traffic received is independent of the mobility of the nodes. In figure 9 we can see that the routing traffic goes from 440 Kbits/s for 250-node case to 250 Kbits/s when there are group of nodes (a 43% of improvement). In the 100-node case, it goes from 230 Kbits/s to 140 Kbits/s when it is a group-based topology (a 39% of improvement). When there are mobility and errors and failures (see figure 10), in the 250-node topology the values go from 440 Kbits/s to 250 Kbits/s in the group-based topology (a 43% of improvement). We obtained 200 Kbits/s in the regular 100-node topology and 135 Kbits/s for the group-based one (a 32% of improvement).

In figure 11, the delay at the application layer simulated for the OLSR protocol using fixed topologies is shown. In the case of 250 nodes we have obtained a delay around 0.015 seconds, which has changed to 0.0035 seconds in the case of 250-nodes group-based topology (there is a 76% of improvement). In the case of 100 nodes, it has decreased from 0.005 seconds in the regular topology to 0.002 seconds in the group-based topology, so there is a 60% of improvement. When there is mobility and errors and failures in the network for the OLSR protocol (see figure 12), we observe that the 100-nodes regular topology has a delay at the application layer of 0.007 seconds when the network has converged, but there is a delay of 0.0025 seconds for the 100-nodes group-based topology (a 64% of improvement). In the case of 250 nodes the improvement is around 60 %. We have obtained a delay of 0.005 seconds in the regular topology versus 0.002 seconds in the group-based topology.

Finally, we have studied the behaviour of the OLSR protocol analyzing the mean routing traffic received (figures 13 and 14). The routing traffic received in the 100-node fixed topology was around 180 Kbits/s, while in group-based topology has decreased to 70 Kbits/s, so there is a 61% of improvement. In the 250-node topology case, we appreciate that this traffic was approximately 300 Kbits/s, but there are values lower than 150 Kbits/s in the group-based topology (figure 13). So there is a 50% of improvement. Figure 14 shows the results of a network with mobility and errors and failures. We have observed some fluctuations due to the failures and errors in the network, in both 100-node and 250-node topologies. Those fluctuations are minimized when we use group-based topologies. Improvements of 61% and 50% are obtained in 100-node and 250-node topologies, respectively.

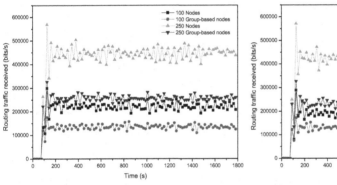

Fig. 9. AODV routing traffic received in fixed topologies.

Fig. 10. AODV routing traffic received in mobile topologies.

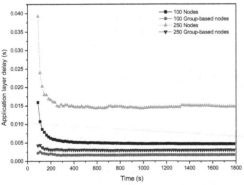
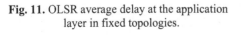
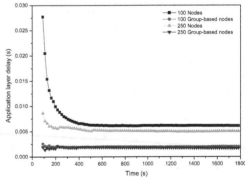

Fig. 11. OLSR average delay at the application layer in fixed topologies.

Fig. 12. OLSR average delay at the application layer in mobile topologies.

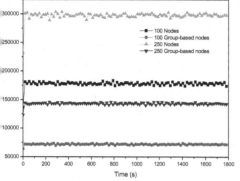

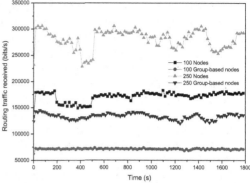

ig. 13. OLSR routing traffic received in fixed topologies.

Fig. 14. OLSR routing traffic received in mobile topologies.

3 Group-based topologies comparison

In order to make the comparison of DSR, AODV and OLSR using group-based topologies, we have used the same test bench used in section 2. This comparison will show us which mobile and ad-hoc routing protocol have better features for group-based topologies.

The average delay at MAC layer in fixed group-based topologies is shown in figure 15. All routing protocols have an average delay lower than 0.001 seconds when the network has converged in both 100-nodes and 250-nodes topologies. It shows that group-based topologies have a good behaviour. DSR protocol with 100-node topology has been the one with worst behaviour and OLSR in 250-node topology has been the best one. OLSR protocol has the same delay (around 0.001 seconds) for both topologies, 100-nodes and 250-nodes, approximately, and it is the most stable. Figure 16 shows the simulation for mobile and errors and failures topologies. All protocols have a delay lower than 0.001 seconds when the network has converged. In this case, AODV protocol has the worst behaviour and OLSR protocol is the most stable.

When the average throughput consumed in the fixed group-based topologies is compared (Figure 17), the protocol that consumes lowest throughput is the DSR protocol (90 Kbits/s in the 100-node topology and 170 Kbits/s in the 250-node topology). The protocol with the most stable throughput consumed is the OLSR protocol. When the network converges, both AODV and OLSR protocols have the same average throughput in the 100-nodes topology, but the OLSR protocol has the lowest convergence time. In case of having a group-based topology with mobility and errors and failures (see figure 18), the results are very similar to the previous ones. The protocol that consumes lower throughput is DSR. AODV protocol consumes lower throughput while the network is converging, but this throughput becomes very similar to the one given by OLSR protocol when the network converges. OLSR protocol is still the most stable.

Then, we analyzed the protocols behaviour when there is MANET traffic. In fixed group-based topologies (see figure 19), the 250-nodes topology shows that the protocol with lower traffic is AODV (40 bits/s approximately) and the one with highest traffic is OLSR. In the 100-nodes topology all protocols have similar behaviour (between 160 bit/s and 180 bits/s). When the network has converged, we can consider AODV and DSR as the best ones and OLSR as the worst. When there is mobility in the group-based topology (see figure 20), the protocol with lowest MANET traffic in the 250-nodes topology is DSR protocol (80 bits/s approximately) and the worst is OLSR protocol. In the case of 100-node topology the one with lowest MANET traffic is DSR protocol and the worst OLSR.

When we analyze the routing traffic sent in fixed group-based topologies (see figure 21) we observe that the one which sends more routing traffic is AODV protocol, (around 120 Kbit/s in the 250-nodes group-based topology and 56 Kbits/s in the 100-nodes group-based topology). OLSR protocol has the best behaviour. It is more stable than the other ones and it sends lower routing traffic than the others (64 Kbits/s in case of the 250-nodes topology and 28 Kbits/s in the 100-nodes topology). When we analyze the mobile group-based topology (figure 22), although the routing traffic has decreased very few, the behaviour of the protocols is very similar to the fixed group-based topologies (Figure 21). AODV is the worst protocol because it is the one which sends more routing traffic to the network and OLSR is the most stable and the one which sends lower routing traffic to the network. The one which has worst stability in mobile group-based topologies is the DSR protocol.

The routing traffic sent is obtained by measuring every node as a source and figures 21 and 22 give the whole routing traffic sent by all of them. However, the routing traffic received is obtained by adding the traffic received by all nodes. The routing traffic received in fixed and mobile group-based topologies is shown in figures 23 and 24 respectively. We can see that it is more than the double of the values obtained for the routing traffic sent. In fixed group-based topologies (see Figure 23) AODV protocol is the one that gives higher routing traffic received (around 250 Kbit/s in 250-nodes topology and 135 Kbits/s in 100-nodes topology). OLSR protocol is the most stable and the one with lower routing traffic received (145 Kbits/s in 250-nodes topology and 70 Kbits/s in 100-nodes topology). When the mobile group-based topologies are analyzed (figure 24), AODV protocol is the one that has worst behaviour and OLSR is the most stable and the one that has lower routing traffic sent. DSR protocol is the most instable.

Figure 25 shows the average delay at application layer in fixed group-based topologies. The protocol most instable and with higher delay in 100-nodes and 250-nodes topologies is AODV protocol. It has peaks with more than 0.45 seconds and it is stabilized around 1700 seconds with a mean value of 0.15 seconds. DSR and OLSR are the ones with lowest delay. Figure 26 shows the average delay at application layer in mobile group-based topologies. DSR protocol is the one that has worst delay till the network converges. Then, when the network is stabilized, the worst is AODV protocol which has delays between 0.1 and 0.15 seconds. OLSR protocol gives the lowest delays.

Then, we have compared DSR and AODV in some common reactive protocols features. In figure 27 the average number of hops in a path for fixed group-based topologies can be observed. DSR protocol has an average value of hops close to 5 in the 250-nodes topology when the network has converged. The number of hops in the 100-nodes topology is slightly lower. AODV has lower average number of hops (around 3.25 hops in the 250-nodes case and 2.75 in the 100-nodes case). The convergence time for the DSR protocol is quite lower than AODV, but it is more instable. In the case of mobile group-based topologies (see figure 28) the behaviour is similar as the previous one, so there is not any dependence on the mobility.

Now we have analyzed the route request sent in reactive protocols for fixed and group-based topologies (figures 29 and 20 respectively). AODV protocol is the one with most number of route requests sent (860 approximately in 250-nodes topology and 330 approximately in 100-nodes topology). We have observed a relationship between the number of route requests sent in the AODV protocol and number of nodes in the topology. There is approximately a factor of 3.3. In the DSR protocol, the number of route requests sent is equal to 730 in the 250-nodes topology and 190 in the 100-nodes topology. Both, fixed and mobile, present the same behaviour. We have observed that the route request sent is the only parameter that gives worst values in group-based topologies than in regular topologies.

Table 1 shows the best and worst protocols for all parameters analyzed.

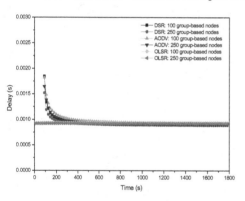

Fig. 15. Comparison of average delays at MAC layer in fixed topologies.

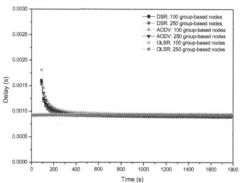

Fig. 16. Comparison of average delays at MAC layer in mobile topologies.

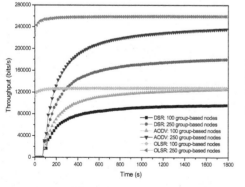

Fig. 17. Comparison of average throughputs consumed in fixed topologies.

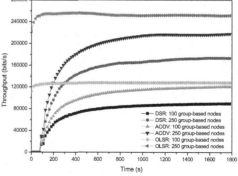

Fig. 18. Comparison of average throughputs consumed in mobile topologies.

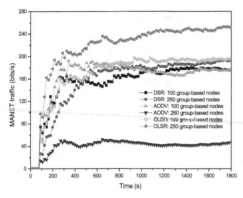

Fig. 19. Comparison of average MANET traffic in fixed topologies.

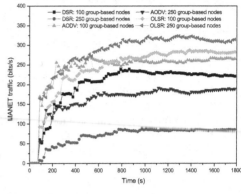

Fig. 20. Comparison of average MANET traffic in mobile topologies.

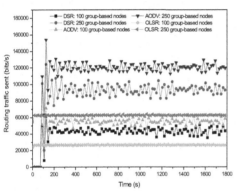

Fig. 21. Comparison of routing traffic sent in fixed topologies.

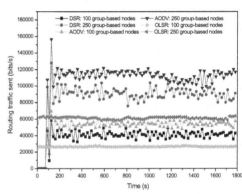

Fig. 22. Comparison of routing traffic sent in mobile topologies.

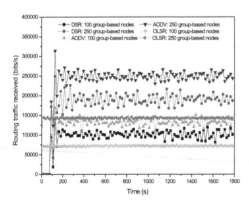

Fig. 23. Comparison of routing traffic received in fixed topologies.

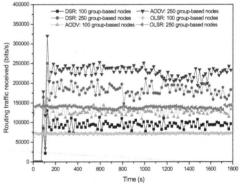

Fig. 24. Comparison of routing traffic received in mobile topologies.

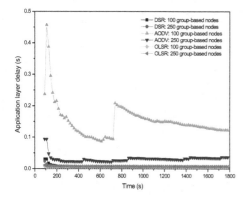

Fig. 25. Comparison of the average delay at application layer in fixed topologies.

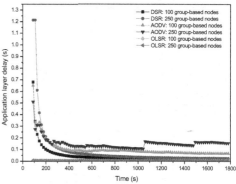

Fig. 26. Comparison of the average delay at application layer in mobile topologies.

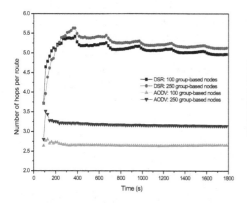

Fig. 27. Comparison of the average number of hops in a path in fixed topologies.

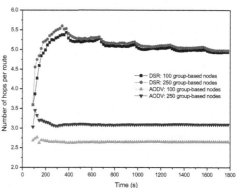

Fig. 28. Comparison of the average number of hops in a path in mobile topologies.

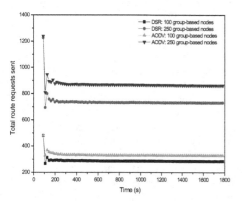

Fig. 29. Comparison of the average number of route requests sent in fixed topologies.

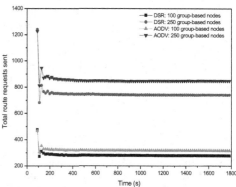

Fig. 30. Comparison of the average number of route requests sent in mobile topologies.

Table 2. Comparison of mobile and ad-hoc routing protocols in group-based topologies.

	Best in fixed	Best in mobile	Worst in fixed	Worst in mobile
Delay at MAC layer	OLSR	OLSR	DSR	AODV
Throughput consumed	DSR	DSR	AODV & OLSR	AODV & OLSR
MANET traffic	AODV	DSR	OLSR	OLSR
Routing traffic sent	OLSR	OLSR	AODV	AODV
Routing traffic received	OLSR	OLSR	AODV	AODV
Delay at application layer	DSR & OLSR	OLSR	AODV	AODV
Average number of hops in a path	AODV	DSR	AODV	DSR
Route requests sent	DSR	AODV	DSR	AODV

4   Conclusions

We have simulated 3 MANET routing protocols with grouping nodes, to demonstrate that group-based topologies improve the network performance. The best improvement percentage has been the DSR protocol when the average delay at the application layer has been simulated. We have observed more improvement in fixed topologies when there are 250 nodes in the topology, but when there is a mobile topology, the improvement is higher in the topology with 100 nodes. When a routing protocol is the best one in a fixed group-based topology, it continues being the best one in the mobile group-based topology. On the other hand, we have observed that a routing protocol, which is the best (or worst) in a group-based fixed topology, could not be the best (or worst) in the mobile topology. The routing protocol that has appeared more as the best one has been OLSR and the one that has appeared as the worst one has been AODV.

References

1. D. Johnson, Y. Hu, D. Maltz. "The Dynamic Source Routing Protocol (DSR) for Mobile Ad-hoc Networks for IPv4". RFC 4728. February, 2007.
2. C. Perkins, E. Belding-Royer and S. Das, "Ad Hoc On-Demand Distance Vector (AODV) Routing". RFC 3561. July, 2003.
3. T. Clausen and P. Jacquet. "Optimized Link State Routing Protocol". RFC 3626. Oct. 2003.
4. R. Misra, C. R. Mandal, "Performance comparison of AODV/DSR on-demand routing protocols for ad hoc networks in constrained situation", IEEE International Conference on Personal Wireless Communications, 2005. Pp. 86- 89. January 2005
5. G. Jayakumar, G. Ganapathy "Performance Comparison of Mobile ad-hoc Network Routing Protocol". Int. Journal of Computer Science and Network Security, Vol.7 No.11, Nov. 2007.
6. H. Jiang and J. J. Garcia-Luna-Aceves, "Performance comparison of three routing protocols for ad hoc networks". Proceedings of IEEE ICCCN 2001.
7. S. Ahmed, M. S. Alam, "Performance Evaluation of Important Ad Hoc Network Protocols". EURASIP Journal on Wireless Communications and Networking. Vol. 2006. Pages 1–11.
8. P. Johansson, T. Larsson, N. Hedman, B. Mielczarek, and M. Degermark, "Scenario-Based Performance Analysis of Routing Protocols for Mobile Ad Hoc Networks". Proc. of the 5th ACM/IEEE int. conf. on Mobile computing and networking. Pp. 195-206. 1999.
9. J. Broch, D. A. Maltz, D. B. Johnson, Y. Hu, J. Jetcheva, "A Performance Comparison of Multi-Hop Wireless Ad Hoc Network Routing Protocols". Proc. of the 4th ACM/IEEE int. conf. on Mobile computing and networking. Pp. 85-97. 1998.
10. Azzedine Boukerche, "Performance Evaluation of Routing Protocols for Ad Hoc Wireless Networks", Journal of Mobile Networks and Applications. Vol. 9, No 4. Pp. 333-342, 2004.
11. OPNET Modeler website. At, http://www.opnet.com/solutions/network_rd/modeler.html

MAC specifications for a WPAN allowing both energy saving and guaranteed delay
Part A: MaCARI: a synchronized tree-based MAC protocol

Gérard Chalhoub, Alexandre Guitton, Michel Misson
{chalhoub,guitton,misson}@sancy.univ-bpclermont.fr

LIMOS-CNRS, Clermont University
Complexe scientifique des Cézeaux,
63177 Aubière cedex, France

Abstract. Industrials have been increasingly interested in sensor and actuator networks to monitor and control installations. The recent IEEE 802.15.4 standard has been developed to address vital issues of these networks, such as limited battery power and low processing capabilities. However, the standard does not meet all the requirements of industrial networks. For example, only some of the IEEE 802.15.4 nodes save energy, and the delay for the computer running the monitoring application to retrieve the sensor data or to activate an actuator is not bounded. Our research on energy-efficient MAC protocol is divided into two parts: Part A is the proposal of a flexible, synchronized tree-based MAC protocol called MaCARI and Part B deals with optimizations that can be performed within each cell.
This paper focuses on Part A, that is, on the description of the MaCARI protocol. MaCARI is designed to tolerate scheduled activities such as sensor data retrieval and unscheduled activities such as complex routing. MaCARI achieves this flexibility by using a tree-based centralized mechanism. We show the benefits of MaCARI by ensuring all nodes sleep regularly and by proving that the maximum end-to-end delay is bounded.

Keywords: wireless sensor networks, IEEE 802.15.4, tree-based synchronization, energy efficient MAC.

1 Introduction

With the advances in electronics, it is possible to build small, cheap, battery-powered devices that can perform basic computations, sense the environment and communicate in a wireless manner. Ideally, these devices could be deployed at a low cost and organize themselves to form a network that monitors an area of interest.

Recently, many research groups and industrials have focused on such wireless networks of sensors and actuators. Since these devices are often battery

Please use the following format when citing this chapter:

Chalhoub, G., Guitton, A. and Misson, M., 2008, in IFIP International Federation for Information Processing, Volume 264; Wireless Sensor and Actor Networks II; Ali Miri; (Boston: Springer), pp. 221–232.

powered, it is a vital issue to reduce the energy consumption of all the network elements (see [1] for a comparison on energy-efficient MAC protocols). The IEEE 802.15.4 standard [2] has been developed to address this problem, and has been implemented on real sensors and actuators.

The OCARI project [3] is a joint project with industrial and academic partners[1], which goal is to develop and study protocols that can increase the lifetime of a sensor and actuator network. The main scenario considered in the project is the monitoring of an industrial environment such as a factory or a production site. Such a network has the following characteristics: (i) it consists of no more than 200 sensors and actuators, with low mobility (although most of the nodes of the network are static, the propagation conditions in the environment constantly change); (ii) every network element has a limited battery-power; (iii) communications between sensors (e.g., a temperature sensor), the monitoring computer (which often is the decision maker) and actuators (e.g., an alarm bell) have a bounded delay. In the context of the OCARI project, we developed a protocol called MaCARI (MAC protocol for Ad-hoc Industrial Networks).

In this paper, we describe the MaCARI protocol. In a nutshell, MaCARI divides the time into three periods forming a global cycle: (i) the *synchronization period* allows all the network elements to be synchronized; (ii) the *scheduled activities period* is dedicated to retrieving the values of the sensors and relaying commands to the actuators; (iii) the *unscheduled activities period* can be used for running sophisticated routing protocols or simply sleeping. These three periods are shown on Fig. 1. In order to obtain the scheduling, we use a centralized approach: a specific node is in charge of creating a tree that spans all the nodes of the network. Synchronization beacons are periodically broadcasted along this tree.

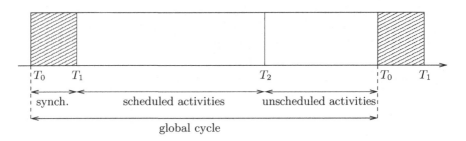

Fig. 1. MaCARI divides the time into three periods: a synchronization period between T_0 and T_1; a scheduled activities period from T_1 to T_2; an unscheduled activities period between T_2 and the next T_0. The first two periods constitute the tree-based activities. The unscheduled activities are non tree-based activities and can be used for any routing protocol.

---

[1] The OCARI project is a partnership between EDF R&D, DCNS, INRIA, LRI, LIMOS, One RF Technology and LATTIS with the support of the ANR (French National Research Agency).

Our contributions are three-fold. First, we merge the two solutions proposed in the 802.15.4 standard in order to avoid direct and indirect beacon collisions. Second, we reduce the energy consumption of the nodes by allowing all devices to sleep. Third, our mechanism implements multi-hop communications with delay guarantees.

This paper is organized as follows: first, we describe the 802.15.4 standard in Sect. 2. Then, we present our protocol MaCARI in Sect. 3. We provide a detailed description of the three periods, emphasizing more on the first two periods. More details on the scheduled activities component are given in Part B of this paper (please refer to [4]). After discussing some open issues in Sect. 4, we conclude our work.

2   IEEE 802.15.4 standard

The IEEE 802.15.4 standard considers two types of devices: end-devices and coordinators that are in charge of the end-devices. The MAC protocol supported in the 802.15.4 standard has two operational modes: the *non beacon-enabled* mode in which beacons are sent on request and no synchronization is required between the coordinators and their children, and the *beacon-enabled* mode in which each coordinator sends periodic beacon to synchronize the activity of its children. Only the beacon-enabled mode allows energy saving, and is therefore suited to our objectives.

2.1   Beacon-enabled mode

In this paper we are interested in the beacon-enabled mode in which the activity of the devices follows the superframe structure shown on Fig. 2. Each superframe starts with a beacon. There is an optional inactivity period between two superframes. The duration of superframes and inactivity periods is specified with the SO (for superframe order) and BO (for beacon order) parameters ($0 \leq SO \leq BO \leq 14$) contained in the beacons:

$$\begin{cases} SD = aBaseSuperframeDuration.2^{SO}, \\ BI = aBaseSuperframeDuration.2^{BO}, \end{cases}$$

where SD is the superframe duration, BI is the beacon interval, and the minimum duration of a superframe is called $aBaseSuperframeDuration$, which is equal to 15.36 ms.

During the superframe the children of a coordinator compete using slotted CSMA/CA, defined in the 802.15.4 standard [2]. An optional mechanism that offers Guaranteed Time Slots (GTS) is proposed by the standard but, to our knowledge, it has not been implemented yet due to its complexity. GTS allow end-devices to ask their coordinator for a time slot during which no other devices can be active in the cluster. All the guaranteed time slots form the contention-free period.

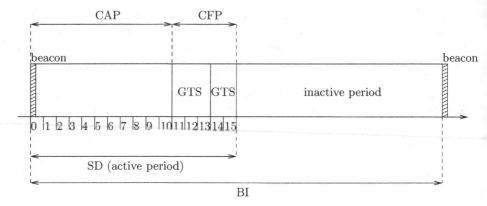

Fig. 2. The 802.15.4 superframe starts with a beacon, is followed by a Contention Access Period (CAP) during which the slotted CSMA/CA is used, and might contain a Contention-Free Period (CFP) based on Guaranteed Time Slots (GTS) allocation. If $BO > SO$, there is an inactive period that comes before the next superframe.

2.2   Cluster tree

The 802.15.4 standard has proposed to interconnect coordinators using a *cluster tree* network, as shown on Fig. 3. A cluster is formed by a coordinator and all its children (end-devices and other coordinators). The cluster tree is the union of all the clusters. A PAN (Personal Area Network) coordinator is in charge of allocating the addresses of the nodes on the tree. Details on how the cluster tree is created can be found in the 802.15.4 standard.

However, the cluster tree network suffers from beacon frame collisions when used in beacon-enabled mode. Two types of beacon frame collisions have been identified by the Task Group 15.4b [5]: *direct* and *indirect* beacon frame collisions. In both cases, two coordinators or more send their beacon frames at approximately the same time. In the first case they are in the same transmission range of each other, while in the second case they cannot hear each other but have overlapping coverage zones. The Task Group 15.4b has proposed two solutions to avoid the direct beacon frame collisions, namely the time division approach and the beacon-only period approach, and two solutions to avoid the indirect beacon frames collisions, namely the reactive approach and the proactive approach.

Solutions to direct beacon frame collisions. In the time division approach, each coordinator has to choose a time interval in a major cycle to schedule its own superframe that does not interfere with the superframes of its neighbors. An implementation for this solution was proposed in [6]. In the beacon-only period approach, all beacons are sent during a beacon-only period which precedes the superframes, each coordinator chooses an empty slot to send its beacon so that no beacon collisions occur.

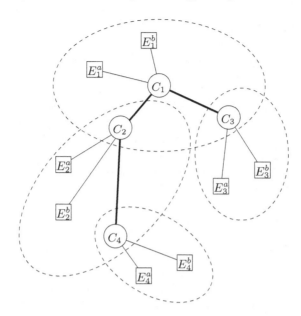

Fig. 3. A cluster-tree is built on a small topology, with C_1 as a PAN coordinator. C_i represents a coordinator and E_i^x represents an end-device associated with coordinator C_i. Every coordinator constitutes a cluster with its children devices. For example, the cluster of coordinator C_2 is formed by $\{C_2, C_4, E_2^a, E_2^b\}$ and the cluster of coordinator C_3 is formed by $\{C_3, E_3^a, E_3^b\}$.

Critics have been made on these two approaches (see [7]):

- In the time division approach: (i) the duty cycles of each cluster are constrained and depend on the number of interfering clusters; (ii) communications between adjacent coordinators is not possible since their duty cycles are separated in time.
- In the beacon-only period approach: (i) the GTS mechanism is no longer possible since all the superframes are scheduled during the same time interval; (ii) dimensioning the duration of the beacon-only period is complex since it depends on the cluster tree, on the number of coordinators and on the allocation of the beacon transmission slots.

Solutions to indirect beacon frame collisions. In the reactive approach, an end-device experiencing beacon frames collisions notify its parent coordinator. This notification is received by other coordinators in range, which change their beacon transmission scheduling accordingly. In the proactive approach, the goal is to avoid indirect beacon frames collisions before they happen. This is achieved during the association procedure, that is, each time a new device joins the network: all the devices in range notify the new device of the beacon transmission time of their parents, in addition to their own beacon sending time for coordinators.

3   MaCARI protocol

In this section, we present in details the MaCARI protocol. As explained in Sect. 1, MaCARI divides the time into three periods that form a global cycle. The three periods are shown on Fig. 1.

Similarly to the 802.15.4 standard, we consider in MaCARI three types of devices: the end-devices, the coordinators and a PAN coordinator. In addition to its original role in the 802.15.4 standard, the PAN coordinator also performs global synchronization, as explained in Sect. 3.1.

MaCARI is a tree-based protocol. It uses the cluster-tree proposed by the 802.15.4 standard. An example of a small network illustrating each cluster is depicted on Fig. 3.

3.1   Synchronization period

The goal of the synchronization period, between T_0 and T_1, is to define the global cycle by providing the same vision of a global time to all the coordinators and end-devices of the tree. This synchronization allows all devices, including the coordinators, to sleep and to wake up at predefined instants, sparing energy while sleeping.

The main difficulty is to broadcast the synchronization in a multi-hop fashion, which increases the error margins on time precision. A beacon is initiated by the PAN coordinator and propagated along the tree by the other coordinators, until it reaches all the devices of the tree.

To make sure that no collisions occur between beacons, the beacon transmission time slot of each coordinator is predefined by the PAN coordinator and included in the beacon itself. Figure 4 shows how beacons are propagated during the synchronization period for the topology shown in Fig. 3.

By T_1, all devices should share the same global time and have their internal clocks synchronized. However, many sources of error affect this synchronization mechanism and have to be taken into consideration [8].

- the processing time before sending the beacon varies with low-level interruptions that schedule the microprocessor activities,
- the processing time after receiving the beacon varies for the same previous reason,
- the propagation delay is dependent on the distances between the devices,
- the clock drift depends on the crystal of each device internal clock.

Three solutions have been proposed by the Task Group 15.4b to reduce the error induced by these factors [5]. These solutions are based on estimating the maximum duration that each source of error might be. They could be easily implemented and included to our protocol, knowing that the distance separating the devices does not exceed 15 meters in our context and therefore the propagation delay can be neglected. The technical details concerning this issue are out of the scope of this paper.

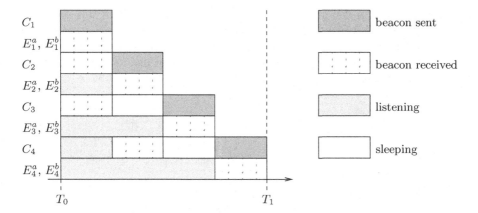

Fig. 4. The synchronization period is a successive transmission of beacons. The PAN coordinator C_1 initiates the beacon propagation with the beacon transmission schedule (C_1, C_2, C_3, C_4). E_1^a, E_1^b, C_2 and C_3 receive the beacon. According to the content of the beacon, C_2 is the next coordinator to propagate the beacon. C_2 sends the beacon, which is received by E_2^a, E_2^b and C_4. Then, C_3 and C_4 propagate the beacon. By T_1, all the devices have received the beacon. The decision to turn to sleeping mode depends on the duration of the waiting time before T_1 or the scheduled beacon transmission. For example, C_1 could decide to sleep until T_1, while C_3 might not have time to sleep before sending its beacon.

3.2 Scheduled activities period

MaCARI schedules the activities of the devices into several activity periods between T_1 and T_2. Each activity period concerns a coordinator and its end-devices, which form a *star*. The star is different from the cluster in that a star contains only one coordinator.

The scheduling of the activity periods of the stars provides a specific activity period to each star without interferences with the other stars (see on Fig. 5). During this period, the coordinator communicates with the end-devices. To allow the coordinators to communicate with one another, the parent coordinator is listening for the entire duration of the activity period of its children coordinators. This creates common active periods between coordinators at the end of the activity period of each child coordinator. The communications during the common active periods are depicted with arrows on Fig. 5. When these two coordinators communicate, all the end-devices of the child star are inactive.

Thus, the activity period of each star is composed of two parts: a first part during which the coordinator collects the data from the sensors or pilots its actuators, and a second part during which it can exchange data with its parent coordinator.

Each coordinator manages the activity period of its star according to the number of its end-devices and their levels of activity. The optional use of GTS is ensured without the risk of collisions caused by communications from other stars.

More details on the intra-star activity is given in Part B of this paper [4], including real measurements on Freescale components [9]. Note that we considered here that all the stars of the network were working during different time intervals; simultaneous activity periods of stars is still possible, but out of the scope of this paper.

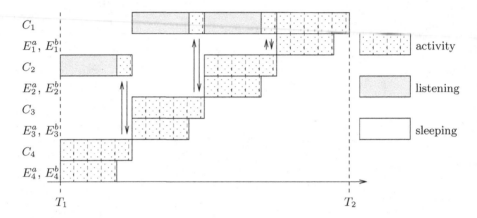

Fig. 5. The scheduled activities period starts at T_1 and ends at T_2. According to the content of the beacon, star of C_4 is the first to be active. At the same time, coordinator C_2 is listening and waiting for C_4 to finish communicating with its children and to initiate a parent-child communication. The same procedure applies to the other stars.

The algorithm used by the PAN coordinator to compute the size of each activity period depends on the traffic load of each star and on the application type. In the example shown on Fig. 5, we assumed that each star has the same traffic load, and the PAN coordinator therefore allocated the same activity period to each star. The same assumption has been made for the parent-child communications. Note that only the traffic with the highest priority uses these parent-child communications. The remaining traffic is forwarded during the unscheduled activities period. As for the application type, we considered that the network consists of more sensors than actuators. Thus, the activities are scheduled from the stars at the bottom of the tree to the stars at the top of the tree (refer to Sect. 4.1 to see how this algorithm can be used to guarantee an end-to-end delay).

3.3   Unscheduled activities period

The period of unscheduled activities, between T_2 and the next T_0, is designed to allow the use of energy-efficient routing protocols. During this period, all the end-devices are sleeping (their have already exchanged data with their coordinators during the period of scheduled activities). MaCARI does not specify the activity of the coordinators: they can either be asleep or active, according to the topology control algorithm [10] used by the routing protocol (refer to [11] for a survey on routing protocols for wireless sensor networks).

The period of unscheduled activities can also be used as a contention access period, where all the coordinators compete for the medium using a stochastic mechanism such as CSMA/CA. Such a mechanism inherently uses simultaneous transmissions, resulting in a good utilization of the channel bandwidth. However, the access to the channel is probabilistic.

The advantage of having scheduled activities in one period and unscheduled activities in another becomes apparent: messages requiring a bounded end-to-end delay are relayed during the period of scheduled activities, according to a path that could be non optimal; other messages can be routed during the period of unscheduled activities using a potentially better path.

3.4   Advantages over the 802.15.4 standard

As explained in Sect. 2, the 802.15.4 standard suffers from direct and indirect beacon frame collisions. In the context of our network specifications, MaCARI solves these problems by:

- merging the two solutions that avoid direct beacon frame collisions,
- avoiding completely the indirect beacon frame collisions.

Unlike the beacon-only period solution:

- The synchronization period (the equivalent in MaCARI of the beacon-only period) is well defined by the PAN coordinator, which knows all the coordinators.
- Since the activity periods of the stars are separated in time, the use of GTS is possible.

Unlike the time division solution, MaCARI allows communications between neighbor coordinators. They can be either parent-child communications happening in the scheduled activities period between T_1 and T_2, or between any pair of coordinators in range during the period of unscheduled activities, between T_2 and T_0. MaCARI does not solve the limitation on the number of clusters of the time division solution. However, this number is limited in the industrial network we focus on. Subsection 4.2 proposes a way to further reduce the impact of this limitation.

In MaCARI, there is no indirect beacon frame collision, since all beacon frame transmissions are predefined and no random choice is made for the sending time of the beacon frames.

In addition, MaCARI allows the coordinators to save their energy by sleeping during certain time intervals. This is not supported by the IEEE 802.15.4 standard.

4 Discussion

In this section, we describe a feature of the MaCARI protocol, namely, how it can be used to guarantee a bounded end-to-end delay. We also discuss how simultaneous transmissions could be used to optimize the synchronization and the scheduled activities periods.

4.1 Guaranteed end-to-end delay

In this part, we explain how high priority data can be relayed from a coordinator to another during the scheduled activities period. Low priority data is relayed during the unscheduled activities period. We prove that, under known traffic conditions, we can guarantee a bounded end-to-end delay. In the context of industrial networks, most sensors have a well defined data production, which is taken into account while planning the scheduled activities.

Let us consider the network topology shown on Fig. 3. Figure 5 represents the following scheduling of star activities (privileging communications from the bottom of the tree to the root of the tree): (C_4, C_3, C_2, C_1), and let us assume that sensor E_4^a (E_j^i represents the i-th end-device of coordinator C_j) has to communicate to C_3. During the activity of star C_4, E_4^a relays its message to C_4. Part B [4] of this paper specifies different optimizations for intra-star activities. C_4 can relay the message to its father C_2 which is listening. Later in the cycle, it is the turn of star C_2 to work. Towards the end of the activity period of star C_2, C_2 can pass the message to its father C_1, which is active. However, in order to communicate to C_3, C_1 has to wait for a new global cycle to start. When it is the turn of star C_3 to work, C_1 can relay the message of E_4^a to its destination C_3 towards the end of the activity period of star C_3.

From the scheduling (C_4, C_3, C_2, C_1), it can be seen that messages can be relayed all the way up the tree to the root in one global cycle. However, each time a message has to go down one level on the tree, the coordinator has to wait for a global cycle.

The fact that messages can be relayed in one global cycle all the way up the tree is due to the scheduling, which always schedules the activity period of child stars before the activity period of the star of their parent. This is called an *upstream scheduling*. The reverse scheduling, which always schedules the activity period of the star of the parent before the activity period of the star of the child is called a *downstream scheduling*, since it allows to go all the way down the tree in one global cycle.

To achieve a bounded end-to-end delay, our idea consists of having an upstream scheduling and a downstream scheduling alternatively. Let us consider the worst-case end-to-end delay from a coordinator C_i to a coordinator C_j. Coordinator C_i has to send the message to the root, with a downstream scheduling for the current cycle. When this cycle ends, the message has reached only the father of C_i. Then, during the second cycle, the scheduling is upstream and the message reaches the root of the tree. Finally, during the third cycle, the scheduling is downstream again and the message can reach C_j. It has taken three global cycles to forward the data from coordinator C_i to coordinator C_j.

This assumes that during the scheduled activities period, the duration of the communication between child and parent coordinators depends on the traffic generated by all the end-devices. This is taken into account by the PAN coordinator during the synchronization period, with the following constraint: each time interval allocated to a star depends on its traffic conditions, and on the position of the coordinator in the tree (coordinators close to the root have to relay more messages).

4.2   Optimizations using simultaneous transmissions

It is possible to reduce the duration of the scheduled activities period by considering simultaneous non-interfering star activities. Simultaneous activities have already been considered for intra-star communications (see the description of SGTS in Part B of this paper). This approach could be extended and applied to realize simultaneous star activities, taking into consideration the received power (and not the transmission range).

It is also possible to reduce the duration of the synchronization period by applying the same strategy. Also, if all the end-devices of a coordinator C_i receive the beacon of another coordinator C_j, C_i does not need to broadcast its own beacon.

5   Conclusions

In this part of the paper, we have presented a flexible, synchronized tree-based MAC protocol called MaCARI. MaCARI divides a global cycle into three time periods: the synchronization, the scheduled activities and the unscheduled activities periods. End-devices and coordinators are both allowed to sleep during these three periods, only to wake up and communicate during predetermined time intervals. Thus, all the network devices are able to save energy.

The period of scheduled activities allows sensors and actuators to communicate with their coordinator. Additionally, during the activity of a child coordinator, the father coordinator is active. This allows guaranteed communications between two adjacent coordinators on the tree, and therefore makes end-to-end communications possible for high priority traffic. By scheduling alternatively upstream and downstream communications, MaCARI is able to relay frames from one end of the network (e.g., a sensor) to another end of the network (e.g., a monitoring application) in three global cycles or less.

The period of unscheduled activities allows the communications between coordinators to be arbitrary (i.e., not respecting the tree structure).

Part A of this paper has focused on the synchronization and on the scheduling of star activities. More details on the optimizations of the communications within each star are given in Part B of this paper [4].

References

1. Halkes, G.P., Van Dam, T., Langendoen, K.G.: Comparing energy-saving MAC protocols for wireless sensor networks. Mobile Networks and Applications (10) (2005) 783–791
2. IEEE 802.15: Part 15.4: Wireless medium access control (MAC) and physical layer (PHY) specifications for low-rate wireless personal area networks (WPANs). Standard 802.15.4 R2006, ANSI/IEEE (2006)
3. The OCARI project: The ocari project web site http://ocari.lri.fr.
4. Livolant, E., Van den Bossche, A., Val, T.: MAC specifications for a WPAN allowing both energy saving and guaranteed delay - Part B: Optimisation of the intra-star exchanges for MaCARI. In: Submitted to WSAN 2008
5. IEEE 802.15 Task Group 4b: TG4b contributions http://grouper.ieee.org/groups/802/15/pub/TG4b.html.
6. Cunha, A., Alves, M., Koubaa, A.: Implementation details of the time division beacon scheduling approach for ZigBee cluster-tree networks. Technical Report TR-070102, Polytechnic Institute of Porto (November 2007)
7. Koubaa, A., Cunha, A., Alves, M.: A time division beacon scheduling mechanism for IEEE 802.15.4/Zigbee cluster-tree wireless sensor networks. Technical Report TR-070401, Polytechnic Institute of Porto (April 2007)
8. Elson, J., Girod, L., Estrin, D.: Fine-grained network time synchronization using reference broadcasts. In: Symposium on Operating Systems Design and Implementation (OSDI). (December 2002)
9. Freescale: Zigbee / IEEE 802.15.4 Freescale solution. Technology General Information, Freescale Semiconductor (2004)
10. Rajaraman, R.: Topology control and routing in ad hoc networks: a survey. ACM SIGACT 33(2) (June 2002) 60–73
11. Al-Karaki, J.N., Kamal, A.E.: Routing techniques in wireless sensor networks: a survey. Wireless Communications 11(6) (2004) 6–28

MAC specifications for a WPAN allowing both energy saving and guaranteed delay [*]
Part B: Optimization of the intra-star exchanges for MaCARI

Erwan Livolant, Adrien van den Bossche, and Thierry Val

Université de Toulouse ; UTM ; LATTIS (LAboratoire Toulousain de Technologie et d'Ingénierie des Systèmes) ; SCSF research group
IUT Blagnac, 1 place Georges Brassens ; BP60073 ; F-31703 Blagnac, France
erwan.livolant@laposte.net, vandenbo@iut-blagnac.fr, val@iut-blagnac.fr

Abstract. Industrials have been increasingly interested in sensor and actuator networks to monitor and control their installations. The recent IEEE 802.15.4 standard has been developed to address vital issues of these networks, such as limited battery power and low processing capabilities. However, the standard does not meet all the requirements of industrial networks. For example, only some of the IEEE 802.15.4 nodes save energy, and the delay for a sensor to activate an actuator is not bounded. Our research on energy-efficient MAC protocol is divided into two parts: Part A consists in introducing a flexible, synchronized tree-based MAC protocol called MaCARI and Part B deals with optimizations that can be performed within each star.
This paper focuses on Part B, that is, on the intra-star MaCARI protocol. Our proposal is an incremental protocol with different options which increase respectively the previous one in terms of bandwidth and energy saving. A hardware prototyping of the last option has been done in order to validate our proposal.
Keywords: wireless sensor networks, IEEE 802.15.4, energy efficiency, QoS, MAC layer.

1 Introduction

The OCARI project [2] aims to optimize the wireless communications for industrial networks with sensors and actuators. One of the challenges of the project is the proposition of a new Medium Access Control (MAC) layer. The physical layer used is the IEEE 802.15.4 PHY2450 [3] also used for ZigBee networks [4]. OCARI network general topology is an ad-hoc mesh network, made of two types of devices: coordinators (routers, i.e. Full Function Devices, or FFD, in

---

[*] The OCARI project is a partnership between EDF R&D, DCNS, INRIA, LRI, LIMOS, One RF Technology and LATTIS with the support of the ANR (French National Research Agency) [1]. LIMOS and LATTIS are in charge of the MaCARI medium access control development.

Please use the following format when citing this chapter:

Livolant, E., van den Bossche, A. and Val, T., 2008, in IFIP International Federation for Information Processing, Volume 264; Wireless Sensor and Actor Networks II; Ali Miri; (Boston: Springer), pp. 233–244.

the IEEE 802.15.4 terminology) and end-devices (sensors, i.e. Reduced Function Devices, or RFD, in the IEEE 802.15.4 terminology). As in the 802.15.4 standard, an end-device must be associated to an unique coordinator. A coordinator and its end-devices form a star topology. However, the coordinators are free to communicate with others coordinators if they are in range with each other (peer-to-peer topology). The general MAC-layer proposed by the OCARI project is named MaCARI (MAC for oCARI) [5]. MaCARI relies on a beaconed tree-based mechanisms between coordinators. This MAC layer organizes the device synchronization and the waking up/transmitting/receiving scheduling. The temporal organization is divided into three parts [5] (Fig. 1).

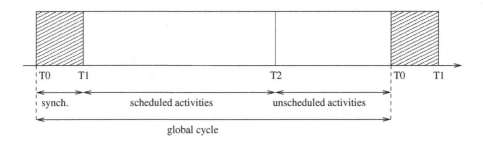

Fig. 1. Global cycle with its three periods.

- A first period, named synchronization period $(T_0 - T_1)$, broadcasts the synchronization and the scheduling over the entire network.
- A second period, named scheduled activities period $(T_1 - T_2)$, is dedicated to the tree-scheduled communications. In this period, each star has a timeslot where no other star can transmit to strictly avoid collisions. The only devices authorized to transmit are the devices of the concerned star (coordinator, end-devices) plus the father coordinator of the star (tree topology). The present article deals with the intra-star organization of this period.
- A third period, named unscheduled activities period $(T_2 - T_0)$, where free communication are possible by using CSMA/CA protocol for instance. In this period, communications may not respect the tree topology. Devices can communicate as soon as they are in the same radio range. The routing protocol proposed by the network-layer uses this period.

2   OCARI basic hypothesis

This section reminds the principal hypothesis of the OCARI project. These hypotheses induce some constraints to MaCARI and have a direct impact on directions and optimizations proposed.

- Each coordinator manages the activity period of its star during the dedicated period it gets between T_1 and T_2. All end-devices are awaken and are waiting for synchronization adjustment. Once the working period has passed, devices may enter in doze mode, as detailed in the next sections.
- The PAN coordinator schedules the use of the medium for each star. This slot enables intra-star communications without collision since the medium is dedicated to a single star during the slot.
- During this period, we suppose that the only star authorized to transmit data is the star which receives the PAN coordinator authorization. In fact, we suppose that collisions are not possible with the others stars.
- Basically there are no direct communications possible between two end-devices. End-devices can communicate with their coordinator thanks to the indirect data transfer described by IEEE 802.15.4 standard [3]. We will see in the next sections that this point can be improved in intra-star MaCARI to optimize throughput.
- The OCARI project assumes a single transmit (TX) power level. A preliminary study [6] has shown that IEEE 802.15.4 transceivers do not save a lot of energy by reducing TX-power (unlike WiFi, WiMAX or GSM transceivers).
- Two same star end-devices may not be in radio range with each others. Of course, a star-coordinator must be in range of all its end-devices.
- Moreover, a star-coordinator may know all its end-devices: the OCARI network is not a spontaneous self-organized network since the application imposes that each sensor is logically "linked" to a particular coordinator. In the OCARI aimed applications, sensors are previously defined and located by the application-level.
- The propagation conditions are quiet stable.
- At last, we consider that devices (end-devices and coordinators) are not mobile. The OCARI network does not permit mobility. At best, end-devices may be mobile but stay in the star, since the protocol does not provide for any handover mechanism.

3   Intra-star MaCARI

Our protocol proposition defines the MAC protocol inside the star when the FFD (Full Function Device) is able to communicate by a beacon reception. During $(T_1 - T_2)$ (Sect. 1), the star owns a guaranteed medium access without collision or interference risks against the other nodes of the OCARI network.

3.1   An incremental protocol: intra-star MaCARI options

We propose several MAC options (Fig. 2), from the simplest to the more complex, as an incremental protocol depending on:

- Comparative performance analysis of these options (by simulation and/or prototype metrology).

- Maximum end-device number by star: indeed, a complex intra-star MaCARI method could be useless if the star owns only a couple of end-devices.
- Quantity and kind of exchanged data: as previously, a complex intra-star MaCARI method could be useless if each end-device only transmits a short frame by minute. On the other hand, if end-devices transmit longer frames with timed-constrained data, with a stronger periodicity, a MaCARI with QoS is required. Such a method, deterministic and with QoS, could enhance the energy saving issue.
- Prototype implementation feasibility and device architecture chose (type of microcontroller(s), memory size, data exchange interfaces, etc.)

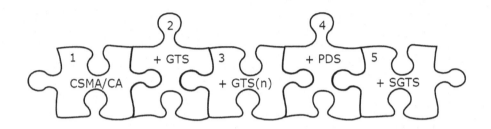

Fig. 2. Incremental MAC options.

3.2   Full slotted CSMA/CA beaconed by the FFD

The simplest intra-star method could be the slotted CSMA/CA one. Every communication is beaconed between T_1 and T_2 during this kind of superframe (as in IEEE 802.15.4 vocabulary). Each node uses the same distributed protocol. As an end-device, a node can only exchange with its FFD. Data for end-devices are transmitted by the FFD during a timeslot requested by the end-device. Concurrent accesses are avoided by the CSMA/CA method. This best effort MAC is relevant for a few-loaded star in regards of the QoS requirements.

3.3   CSMA/CA + GTS (Guaranteed Time Slot)

The random characteristic of the CSMA/CA implies a collision risk between two or more transceivers. In fact, quasi-simultaneous CCAs (Clear Channel Assessment) function calls by concurrent transceivers entail a collision. Without acknowledgment response because of the data collision, each node uses a random backoff time which could be unfortunately the same. In consequences, a new collision might appear. This phenomenon has a small but non-null probability to continue endlessly. The consequences of this could be insignificant for most applications but could be disastrous when the submitted network load is important. Indeed, the more frames to be transmitted, the more collisions, and the more frames to be retransmitted. A collapse phenomenon occurs which implies a quasi-null efficient bandwidth. Delays are also increased. This fact could be unacceptable for time-constrained applications.

This problem could be avoided by using, in each star, a specific access method using a superframe with two periods: CAP (Contention Access Period) and CFP (Contention Free Period). CAP uses only CSMA/CA. This period permits a basic QoS for non-priority flow without temporal guarantees. This best effort algorithm can be used for sporadic and unexpected flows. CFP uses dedicated timeslots allocated by the FFD to its end-devices. These slots are named GTS (Guaranteed Time Slot).

According to the application needs of each sensor, the end-device requests to its coordinator a GTS allocation for the next superframe. If the minimal size of the CAP is not reached, the FFD can allocate a new GTS. The CAP is reduced accordingly. If an end-device does not use its GTS during a specific duration, this GTS is automatically suppressed (timeout). The end-device must request a new GTS attribution for next superframes. The IEEE 802.15.4 standard has proposed this algorithm, but it is rarely implemented on commercial devices.

Two QoS levels for two different flow types are proposed. The first one for time-constrained flows is associated with known needs. The second one for best effort flows is associated with sporadic and unexpected data.

3.4   CSMA/CA + GTS with multiple reservation levels

The previous solution is an interesting option but its major drawback is the GTS static allocation for each superframe. For example, one GTS by superframe could be useless for a simple temperature sensor. It would be interesting to propose a service differentiation according to communication needs of end-device application. This intra-star MaCARI option proposition is based on special reservation level named 'n'. A GTS(n) is dedicated to an end-device according to its periodicity request:

- when n=0: for each superframe (as in IEEE 802.15.4 standard),
- when n=1: every two superframes,
- in the general case: every 2^n superframes.

The principal advantage is the cohabitation of different guaranteed traffics according to different sensors. It is also possible to choose a reverse allocation principle. In a superframe a sensor could have more than one GTS. In this case, the reservation level is n=1. However, the end-device must request several GTSs in each superframe. The bandwidth allocated to such an end-device is increased.

The other advantage of this option is the power saving offered by this MAC layer. A sensor can commute to doze mode, especially when this sensor is not concerned by these superframes. If a temperature sensor has a high inertia, its end-device can wake up only every 4 or 8 superframes for instance for a fast and short temperature transmission. After this activity this end-device commute in battery saving mode. In the classical IEEE 802.15.4 protocol, a wake up is mandatory every superframe for using a GTS. In our proposition, it is possible to save timeslots with an optimized allocation only when it is necessary. This option maximizes doze mode outside GTSs.

3.5   CSMA/CA + GTS(n) + PDS (Previously Dedicated Slot)

The principal drawback of the previous option (also in the IEEE 802.15.4) is related to the GTS request which is necessary done by the end-device in CSMA/CA mode. This request mode is not fully deterministic because this random access is non-guaranteed to obtain GTS which is guaranteed. For sensible sensor applications that need guaranteed access, we propose a new intra-star MaCARI option based on PDS (Previously Dedicated Slot). A PDS is allocated to each specific sensor that uses this high level of QoS. A PDS is in fact an preallocated GTS. This PDS can be used by an end-device to transmit periodic data with high QoS. The end-device can also use this PDS to request more or less GTSs. This dynamic GTS allocation permits variation according to bandwidth and delay needs.

To avoid an overload of the used bandwidth, a large 'n' is associated to this first PDS. A performance analysis [6, 7] shows that the bandwidth used for this PDS is very short (0.78% for a PDS with n=3).

3.6   CSMA/CA + SGTS (Simultaneous GTS)

The last proposed option permits a large-scale OCARI network. In fact, if the network is composed of a large number of devices, particularly for the intra-star part, the congestion is highly probable if the sensors are numerous. The star topology entails a bottleneck inside the range of the coordinator where every data flows converge from end-devices. In some applications, two end-devices could have to exchange data with each other. In this case, a centralize communication via the coordinator can be damageable for the star bandwidth. Typically, a sensor could directly send data to an actuator in the same star, and the actuator could send back an acknowledgment. Both end-devices must be at radio range with each other. This principle optimizes the end-to-end delay and reduces the load of the coordinator.

Moreover, multiple simultaneous communications between end-device pairs in the same star are conceivable under specific radio propagation conditions. Received signal strength of pairs must be different enough in order for a receiver to focus on its associated flow. The SGTS concept [8] consists in allocating the same timeslot to multiple end-device pairs (simultaneous medium access without collisions) (Fig. 3).

- (A) SGTS between end-device pairs
- (B) SGTS between [the coordinator and an end-device] and [an end-device pair]

SGTS feasibility conditions are based on:

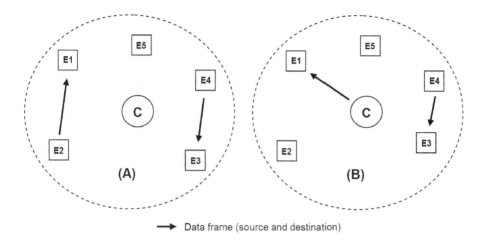

→ Data frame (source and destination)

Fig. 3. SGTS concept.

- A measure of received signal strength from each emitter. This measure must be done in real time for every node which wants to using SGTS.
- A bilateral negotiation between nodes that want a simultaneous timeslot.
- A certain stability of propagation conditions.

For example (Fig. 3(A)) when E2 and E4 transmit data simultaneously, the end-device E1 measures that the signal strength of E4 is lesser than the signal strength of E2 with an additional margin. The coordinator (C) centralizes SGTS-related data and organizes SGTS allocations to its end-devices.

The coordinator can be involved in a pair which negotiates a SGTS allocation. In Fig. 3(B), C communicates with E1 while E4 communicates with E3. In other words, a SGTS is allocated to C⇒E1 while an other SGTS is allocated to E4⇒E3, only if radio conditions are satisfied (as described previously).

As a parameter, the signal strength can be adjusted to increase the number of potential SGTS. In Fig. 4, E4 reduces its signal strength in order to reach only its nearest neighbor E3. In this case, SGTS between E1⇒C and between E4⇒E3 are possible.

The SGTS concept can be used for data frames without acknowledgments. In the case of acknowledgment requirements, acknowledgments must apply the same signal strength measurement and SGTS negotiation procedures. Thus, two different methods can be used:

- Accept a SGTS only if propagation conditions are symmetric.
- Acknowledge data in a specific field of an other data frame associated to a later SGTS.

A theoretical and metrological study on propagation conditions proved an acceptable reception threshold for SGTSs without collision risks. A margin must be respected to increase SGTS quality.

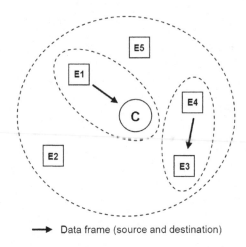

→ Data frame (source and destination)

Fig. 4. SGTS with a receiver FFD.

4 Preliminary study for SGTS concept validation

The concept of SGTS needs to be validated and the non-perturbation margin has to be identified. The best way to validate this concept is the real hardware prototyping. Therefore we have developed a prototype of a simple MAC-layer and deployed a network based on a couple of Freescale IEEE 802.15.4 devices [9]. This type of 802.15.4 devices is totally reprogrammable, which allowed us to implement the SGTS functionality. This prototype also enables us to evaluate the non-perturbation margin evoked in Sect. 3.6.

4.1 The prototype network and its topology

The prototype network is composed of five nodes: a star-coordinator (C) and four end-devices associated (synchronized) to this coordinator. The network topology and the superframe structure (frame scheduling) are respectively shown on Fig. 5 and Fig. 6.

For this case study, we consider that both E1 and E3 have obtained a GTS and can freely use it to send data to another end-device (respectively to E2 in slot #2 and to E4 in slot #3). Slot #4 is used by both E1 and E3. In this study, the objective is to evaluate the non-perturbation margin, i.e. to measure the number of collisions in slot #5. In order to evaluate the perturbation of the other transmitter (E3 for E2 and E1 for E4), each coordinator listens to the messages sent by the two nodes and gets the RSSI value during the slots #2 and #3. The RSSI difference is calculated at the end of slot #3, only if both messages from E1 and E3 have been received. In slot #4, the receivers E2 and E4 listens to the medium. If they receive the expected message, the result is positive. If they receive the message of the other device or if a collision is detected, the result is negative.

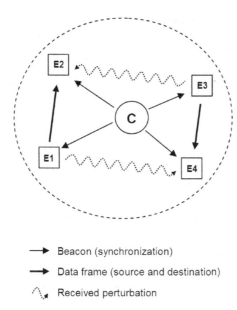

→ Beacon (synchronization)

→ Data frame (source and destination)

⌒⌒◄ Received perturbation

Fig. 5. Prototype topology.

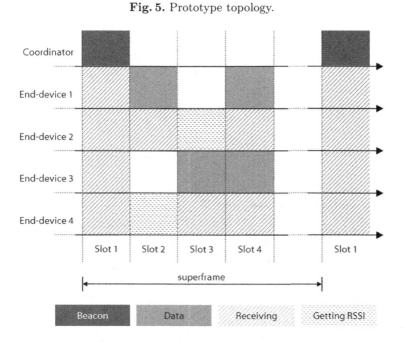

Fig. 6. Message temporal scheduling.

Note that on the devices used for this study, the transmit power can be adjusted from -16 dBm to +3.6 dBm. This functionality enables us to implement an automatic variation of the node transmitting power to increase the measure range without moving the devices. Measures have been realized both into an anechoic chamber (i.e. without any noise) and in a regular environment (i.e. with noise).

4.2   Obtained results

The results showed on Fig. 7 prove that the SGTS concept works correctly on a real prototype: two transmissions can be done at the same time without interfering the other receiver. The signal strength margin is small. In fact, the obtained results show that there is only a 6-7 dB window where the SGTS may not be envisaged because of an important risk of collision. We can considerate that a margin of 10 dB is safe.

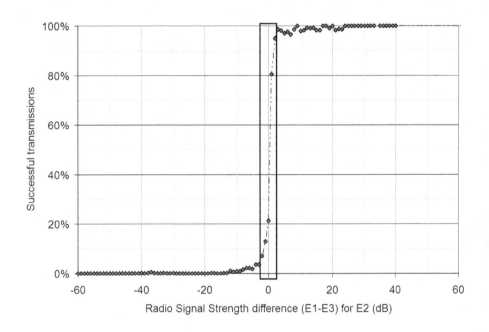

Fig. 7. Obtained results with the SGTS study in regular environment.

5   Conclusion

MaCARI proposes a core based on beaconed slotted CSMA/CA and allows several optional functionalities. The Figure 8 summarizes the five proposed options with their advantages and drawbacks. We plan to analyze the performance of these additional functionalities in a real prototype and consider the results as a reference. The option #2 (IEEE 802.15.4 GTSs) may propose a first level of determinism. Unfortunately, there is no commercial implementation of the GTS. This specific part has to be implemented first. The option #3, which requires the option #2, proposes a higher QoS flexibility thanks to a mechanism of service differentiation. Moreover, the energy-saving resulting of this option was previously required by the OCARI project which aims to propose a MAC-layer with energy-saving functionalities. Option #4 which is based on PDS also requires options #2 and #3. It enables the higher level of determinism. Option #5 plans to optimize the medium utilization between the network entities. In fact, the 250kbps has to be shared in the best way. This last option may be implemented without option #2, #3 and #4. Nevertheless the SGTS concept is very interesting if it is used together with GTS(n) and PDS.

	IEEE standard	Our Contribution	Comments
① CSMA/CA	✓		*Simplest but only best effort*
② CSMA/CA + GTS	✓		*QoS but generally unimplemented*
③ CSMA/CA + GTS(n)		✓	*Service Differentiation*
④ CSMA/CA + GTS(n) + PDS		✓	*Fully deterministic*
⑤ CSMA/CA + SGTS		✓	*Bandwidth optimization*

Fig. 8. Summary of the MAC options.

References

1. ANR: The   ANR   web   site   http://www.agence-nationale-recherche.fr/documents/aap/2006/finance/Telecom-2006-resumes.pdf.
2. The OCARI project: The OCARI project web site http://ocari.lri.fr.
3. IEEE 802.15: Part 15.4: Wireless medium access control (MAC) and physical layer (PHY) specifications for low-rate wireless personal area networks (WPANs). Standard 802.15.4 R2006, ANSI/IEEE (2006)
4. Zigbee Alliance: Zigbee Specification. Standard Zigbee R2005 (2005)
5. Chalhoub, G., Guitton, A., Misson, M.: MAC specifications for a WPAN allowing both energy saving and guaranted delay - Part A: MaCARI: a synchronized tree-based mac protocol. In: Submitted to WSAN 2008
6. van den Bossche, A., Val, T., Campo, E.: Prototyping and performance analysis of a QoS MAC-layer for industrial wireless network. In: FET2007 (7th IFAC International Conference on Fieldbuses and nETworks in industrial and embedded systems). (2007)

7. van den Bossche, A.: Proposition d'une nouvelle méthode d'accès déterministe pour un réseau personnel sans fil à fortes contraintes temporelles. PhD thesis, Université de Toulouse (2007) In french.

8. van den Bossche, A., Val, T., Campo, E.: Proposition and validation of an original MAC layer with simultaneous medium accesses for low latency wireless control/command applications. In: IFAC WC08 (17th IFAC World Congress). (2008)

9. Freescale Semiconductor: Zigbee / IEEE 802.15.4 Freescale solution. Technology General Information (2004)

ERFS: Enhanced RSSI value Filtering Schema for Localization in Wireless Sensor Networks

Seung-chan Shin and Byung-rak Son and Won-geun Kim and Jung-gyu Kim

Department of Information Communication Engineering, Daegu University, Republic of Korea
{ scshin,brson,wgkim,jgkim }@daegu.ac.kr

Abstract. In this research, we have suggested the Localization Algorithm using Probable Filtering Schema of RSSI without additional hardwares. The existing method has been filtering with only average and feedback of received RSSI values. This method was not considering about the variation of RSSI when obstacles are moving at indoor environment. In this research, we have suggested the probable filtering algorithm which is considered factors of errors at indoor environment and we have demonstrated the superiority of this algorithm through the examination. It presents 14.66% and 11.65% improved accuracy than the existing filtering algorithm.

1 Introduction

The localization technology is one of the key technologies to realize invisible technology in the ubiquitous society. It is possible to make an active computing environment by automatic sensing. It is also possible to give users the useful information naturally without any recognition. Therefore, there are many localization technologies which has been studying with various communication method.

Especially, LBS(Location Based Service), which used GPS(Global Positioning System) with the auto navigation system and mobile network based, is already used in various fields and it has been developing indispensably[1].

Sensor network, one of the key technologies on ubiquitous computing, has been studied with the development of ubiquitous computing technologies. The node localization system has been particularly studied in various application fields for developing technology and reducing errors of measurement. The applications which are based on the local information of nodes are Home Automation, Preventing missing child system, Preventing stranger system, Chasing the location of patient and doctor at the hospital, Analysis of consumer preference at big market or department store, Preventing disaster system and so on[4][5][6][7][8].

Generally, indoor localization system uses the RSSI(Received Signal Strength Indication) with ZigBee protocol in wireless sensor network. Because there are some advantages. It is possible to localize without additional hardwares. It is cheaper for setting system, and it has wide application range. However, there are also some disadvantages on RF such as diffraction, reflection, multi-path, and so on.

The localization system with RSSI could measure almost accuracy at indoor environment where could get LoS(Line of Sight), while the RSSI values are not correct at Non-LoS environment because of obstacles.

Please use the following format when citing this chapter:

Shin, S-c., et al., 2008, in IFIP International Federation for Information Processing, Volume 264; Wireless Sensor and Actor Networks II; Ali Miri; (Boston: Springer), pp. 245–256.

In this research, we have suggested the RSSI filtering algorithm to reduce errors when we localize at indoor environment and another filtering algorithm to accurately localize without additional hardwares. We have been struggling to reduce errors of RSSI at indoor environment because of obstacles by using those filtering algorithms.

2   Related research

2.1   General process of localization

There are five steps on processing of localization. Those are collecting of the location information, converting of the location information, filtering, calculating of the location value, and smoothing. Among these steps, it must involve collecting of the local information, converting of the location information, and calculating of the location value. There are processes about each step[9][10][11][12].

The collecting of the location information is performed between terminal node and beacon node. The local information could be RSSI, ToA(Time of Arrival), or AoA(Angle of Arrival). The variation of the location information is one of processes to use on calculating of the location value. It could be converted using propagation model into distance in terms of location information types. If the location information is RSSI, it can use Friis's propagation model. If it is ToA, it can use the propagation model which is based on physics related with light of propagation.

Filtering is the process which is selecting the changed distance of the location information for location calculating more accurately. There are Cell-ID, ToA, TDoA(Time Difference of Arrival), AoA, and Fingerprint in the method of location calculating. Smoothing is the process which reflects real-time location information, using a location value of nodes which has been received before[2][3].

2.2   Characteristics of RSSI value

Generally, RSSI values are presented as equation 1 at outdoor environment where it is guaranteed LoS. If the equation 1 is presented by graph on ideal environment. It will be same as figure 1[13].

$$RSSI = -(10nlog_{10}d + A) \qquad (1)$$

- n : signal propagation constant
- d : distance from receiver
- A : RSSI at indoor environment far from 1m

It presents the RSSI values at outdoor environment in 1m on figure 2. We could recognize that it presents the distribution regularly, and it is distributed by bisymmetry based on -11dBm.

However, there are great differences in received RSSI values at indoor environment because of obstacles. The figure 3 presents the variation of RSSI values at outdoor environment in 1m. We could recognize that the RSSI values are not bisymmetry.

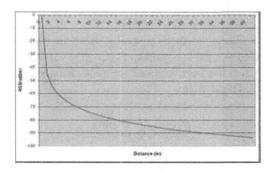

Fig. 1. Ideal RSSI values (on equation 1, A=40, n=3)

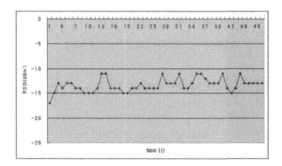

Fig. 2. The variation of RSSI values in terms of received time at outdoor environment in 1m.

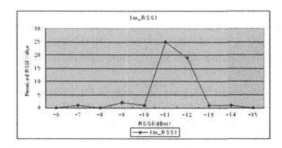

Fig. 3. The frequency of distribution curve of RSSI values at outdoor environment in 1m.

2.3   General filtering method

There are two ways of filtering methods generally which are average and feedback.

The average filtering method is presented as equation (2). It has variable formation that the RSSI value which has been received before is changed by another RSSI value which has been received right after.

$$\overline{RSSI}_n = \frac{1}{n} \sum_{i=0}^{n} RSSI_i \tag{2}$$

- n : the number of received RSSI value
- $RSSI_n$: the received RSSI value in round n
- $RSSI_i$: the received RSSI value in round i

The feedback filtering method is presented as equation (3). It has the variable formation with the RSSI value which has received before exchanges another RSSI value which received right before.

$$RSSI = a \cdot RSSI_n + (1 - a) \cdot RSSI_{n-1} \tag{3}$$

- $RSSI_{n-1}$: the received RSSI value in round n-1
- a : weigh constant$(0 < a < 1, generally, a \geq 0.75)$

The average and feedback filtering method could be available at outdoor environment when the LoS is guaranteed but, it has some problem at indoor environment which illustrate lower accuracy because of obstacles environment which illustrate lower accuracy because of obstacles.

3   Suggested filtering method

The RSSI value can not be measured by obstacles accurately at indoor environment. If the RSSI value which has lower accuracy uses in general filtering methods; average or feedback filtering method, it could make big errors. In this research, we have suggested the enhanced probable filtering method to obtain a higher accuracy than average or feedback filtering method.

In figure 4, it illustrates to select the k_{max} among the measured RSSI values on the suggested filtering algorithm. It can be separated into area A and area B. It could be the higher accuracy RSSI value between transmitter node and beacon node in area A. But, it could be received the lower accuracy RSSI value in area B because there is an error by obstacles. The problems could improve the accuracy of RSSI values by removing the error tolerance by obstacles and getting the average.

If there are errors by obstacles at indoor environment as figure 4, the average and feedback filtering algorithm present big difference compared with ideal RSSI values in figure 2. In this research, we have examined to get higher accuracy RSSI values at

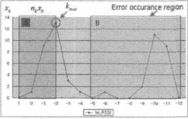

- x_k : the number of received accumulation
- n_k : the range of available error rate
- k_{max} : the maximum frequency of received RSSI value
- A : the range of reliability
- B : the area of occurring error

Fig. 4. The variation of RSSI values at indoor environment in 1m; it illustrates the RSSI values with k_{max} of suggested filtering method

indoor environment with previous RSSI value of k_{max} and to remove the scattered form of the area B in an accumulated distribution chart as figure 4.

The processes of execution are the first, to check the frequency table with received RSSI values, second, to figure out the maximum constant, at last, to calculate the average from the maximum constant to the highest k_{max}. It comes under area A in figure 4. The equation 4 illustrated the suggested filtering algorithm. The figure 5 is a flow chart of suggested filtering algorithm.

$$RSSI_k = \frac{\sum_{k=1}^{k_{max}} n_k x_k}{N_{k_{max}}} \qquad (4)$$

- x_k : Data(RSSI value)
- n_k : times
- $n_k x_k$: Data * times
- N_k : cumulative times
- $k_m ax$: maximum times

There is a disadvantage which keeps the frequency table on memory compared with average filtering method but, it does not increase the complexity. The frequency table doesn't matter to operate sensor nodes because it is stored small amount of memory. For example, if a frequency table has 50 RSSI values, it must need only approximately 50bytes.

4   Examination environment and result

4.1   Examination environment

We have examined the environment in corridor as figure 6. We considered some obstacles such as walls and appliances. We set the sensor node on the 1.5m fixed body from

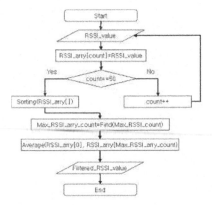

Fig. 5. The flow chart of suggested filtering g algorithm

bottom. We repeated comparing and analyzing the received packets 50 times over 15m because of the available communication range of ZigBee.

Fig. 6. Examination environment

4.2   Hardware and Firmware

We used the Nano-24 made by Octacom. co., ltd. These are sensor nodes which are based on Nano-Q+ and it's developed for education and development kit. MCU is ATmega128L used RISC structure. It supports an interior flash memory based on ISR(In-System Reprogrammable), a 4Kbytes SRAM and a 4Kbytes EEPROM. It also supports an exterior 512Kbytes flash memory and a 32Kbytes SDRAM. Nano-24 is made up of four modules such as Main, Interface, Sensor and Actuator.Figure 7 show structure of main module Among these, we used the main module which involves ATmega128L and CC2420 as figure 8[14][15].

We did firmware porting on sensor nodes which we have developed except OS for sensor node. The figure 8 presents the firmware structure. We optimized that it doesn't

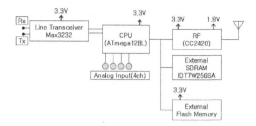

Fig. 7. Structure of main module

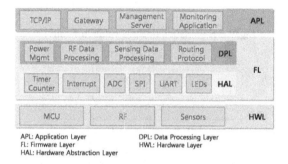

APL: Application Layer   DPL: Data Processing Layer
FL: Firmware Layer   HWL: Hardware Layer
HAL: Hardware Abstraction Layer

Fig. 8. The structure of Firmware

depend on the OS. We also reduced the MCU resource consumption than TinyOS. For example, we reduced flash memory about two times, SRAM about 1.45 times than CntToLedAndRfm which is a RF LED testing program. We actually could reduce the resource consumption which is a simple RF testing program without data if we regard to CntToLedAndRfm. The efficiency of memory management could realize more flexibly on transport layer and network layer with minimizing hardwares in further future.

4.3   Result of examination

The figure 9-11 present the received RSSI values on each time with suggested algorithm from 1m-15m ideal values as it is farther and farther. It causes by obstacles such as walls and appliances at indoor environment.

The figure 9-11 are the graph which is represented table 1 and 2. It presents the comparing result with three kinds of filtering methods. The distance is a standard with ideal RSSI values and it compares with the error of distance. The negative means that the suggested filtering algorithm has lower accuracy, while the positive means that it has higher accuracy in improving rate. It has similar accuracy with feedback filtering method or more without in 2m. The reason is that it presented big improvement of error rates because the last one has slight different value among received 50 RSSI values in 2m. It is the special case. In other case, it causes big errors than other methods. The algorithm presents the highest accuracy in 4m.

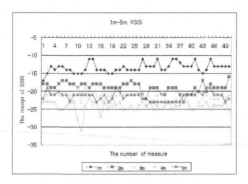

Fig. 9. The variation of RSSI value in 1-5m

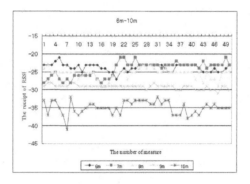

Fig. 10. The variation of RSSI value in 6-10m

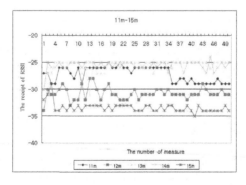

Fig. 11. The variation of RSSI value in 11-15m

Table 1. Comparison of received RSSI value, Ideal, ERFS, Average and Feedback Filtering scheme

Ideal		ERFS	Average Filtering	Feedback Filtering
RSSI	distance	RSSI	RSSI	RSSI
-5.00	1	-13.00	-13.36	-12.99
-11.02	2	-18.35	-18.90	-17.19
-14.54	3	-20.83	-21.58	-21.30
-17.04	4	-22.78	-24.08	-24.36
-18.97	5	-20.63	-21.44	-22.15
-20.56	6	-22.88	-23.72	-24.35
-21.90	7	-22.64	-24.24	-21.80
-23.06	8	-24.70	-25.42	-25.49
-24.08	9	-28.84	-29.14	-29.16
-25.00	10	-33.94	-34.90	-34.95
-25.82	11	-26.34	-27.14	-28.96
-26.58	12	-30.35	-30.78	-30.89
-27.27	13	-25.00	-25.58	-25
-27.92	14	-24.96	-25.60	-25.30
-28.52	15	-33.34	-33.38	-33.75

Table 2. Comparison of conversion distanc, Ideal, ERFS, Average and Feedback Filtering scheme and accuracy comparison of each other

Ideal		ERFS		Average Filtering		Feedback Filtering		Advance	
RSSI	distance	conv. dis	conv. err	conv. dis	conv. err	conv. dis	conv. err	P/A(%)	P/F(%)
-5.00	1	2.51	1.51	2.33	1.61	2.50	1.50	10.62	-0.22
-11.02	2	4.65	2.65	4.41	2.95	4.07	2.07	15.19	**-28.95**
-14.54	3	6.19	3.19	6.01	3.74	6.53	3.53	18.39	11.30
-17.04	4	7.75	3.75	8.01	4.99	9.29	5.29	**31.04**	**38.51**
-18.97	5	6.04	1.04	5.91	1.63	7.13	2.13	11.77	21.63
-20.56	6	7.83	1.83	7.69	2.62	8.45	2.45	13.18	10.29
-21.90	7	7.62	0.62	8.16	2.16	7.51	0.51	21.98	-1.50
-23.06	8	9.66	1.66	9.35	2.49	10.23	2.23	10.39	7.11
-24.08	9	15.99	6.99	14.35	7.06	17.17	8.17	0.78	13.11
-25.00	10	27.99	17.99	27.86	21.26	31.56	21.56	**32.61**	35.63
-25.82	11	11.67	0.67	12.79	1.79	15.77	4.77	10.19	37.33
-26.58	12	18.53	6.53	19.45	7.45	19.72	7.72	7.67	9.91
-27.27	13	10.00	3.00	10.69	2.30	10	2.99	-5.31	-0.05
-27.92	14	9.95	4.04	10.71	3.28	10.36	3.63	-5.41	-2.88
-28.52	15	26.14	11.14	26.24	11.24	27.41	12.41	0.66	8.48
P:ERFS / A:average filtering / F:Feedback filtering / conv.:conversion						average adv		**11.58**	**10.65**

It is the graph which compares with filtered RSSI values. We could recognize that the suggested algorithm is closer to the ideal RSSI value than average and feedback filtering method.

In the figure 12, 13, when the algorithm is applied to the ideal RSSI value and distance, we could find out improved accuracy because there are few conversion distance errors than average and feedback filtering method.

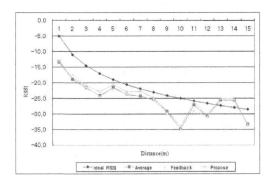

Fig. 12. Comparing the suggested RSSI value

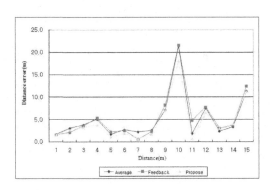

Fig. 13. Comparing distance error with converted RSSI

Lastly, when it compares with the average filtering method, it presents 32.16% improved accuracy in 10m. It also presents 38.51% improved accuracy in 4m when it compares with the feedback filtering method. It has 11.58% and 10.65% average improvement on each average and feedback filtering method.

5   Conclusion

In this research, we have suggested RSSI filtering algorithm without additional hardwares. It improved an accuracy in distance estimation compared with existing average and feedback filtering algorithm on the distance estimation system. Especially, we have examined for the improvement of localization accuracy at indoor environment. It is simple so we could improve the localization accuracy on micromini sensor with low electronic power without additional calculation and complexity. We have demonstrated how much it has been improved. We recognized that the result had improved the 11.58% accuracy than average filtering method, 10.64% accuracy than feedback filtering method. However, we didn't consider about moving nodes. In addition, there is a disadvantage increasing the packet between nodes for distance estimation. After this time, we could study to improve the distance estimation accuracy, considering moving nodes and minimized increasing rate of packets.

Acknowledgement: This research was financially supported by the Ministry of Commerce, Industry and Energy (MOCIE) and Korea Industrial Technology Foundation (KOTEF) through the Human Resource Training Project for Regional Innovation and in part by MIC & IITA (07-Infrastructure-10, Ubiquitous Technology Research Center).

References

1. Lee, Yang, Lee, Cha.: Localization Technology in Unbiquitous Environment. Korean Society for Internet Information, Vol 7. No 2, pp.30–37 (2006)
2. Hakyoung Kim: Location Information Service based Wireless Lan. Telecommunications Review, Vol 16, pp.188–202, (2006)
3. P. Bahl, V.N. Padmanabhan,: RADAR : An In Building RF-Based User Location and Tracking System. In Proceeding of IEEE infocom 2000 Conference on Cmputer Communication, Vol.2, pp.775–784 (2000)
4. Ekahau, Inc., http://www.ekahau.com
5. Place Lab at Intel Corporation, http://www.placelab.org
6. R. Want, A. Hopper, V. Falcao, and J. Gibbons.: The Active Badge Location System. ACM Trans. on Information Systems, Vol.10, pp.91–102 (1992)
7. N. Priyantha, A. Chakraborty, and H. Balakrishnan: The Cricket Location-Support System. Proc. of the ACM Int'l Conf. on MobICom, (2000)
8. J. Hightower, R. Wand, and G. Borriello: Spoton: An Indoor 3d Location Sensing Technology Based on RF Signal Strength. Technical Report 00-02-02. University of Washington (2000)
9. SpotON: Ad-hoc Location Sensing, http://portolano.cs.washington.edu/projects/spoton
10. Portolano: An Expedition into Invisible Computing, http://portolano.cs.washington.edu
11. Hakyoung Kim, Sungduk Kim, Donggil Sue, Jungkang Ji, Hyuntae Jang: A close distance localization Technology tendency. IITA weekly technology tendency, No. 1322, pp.1–12 (2007)
12. Wonhee Lee, Wooyoung Lee, Minkyu Kim, Duseup Eum, Jinwon Kim: Location Measurement System Technology tendency for Ubiquitous Environment. The Korean Institute of Information Scientists and Engineers, Vol 22. No 12, pp.41–50 (2004)

13. K. Aamodt: CC2431 Location Engine. Applications Note AN042. Texas Instrument Incorporated (2006)
14. Octacomm Inc.: http://www.octacomm.net
15. Octacomm Inc.: Understanding of Embbeded System, (2006)

Energy-Efficient Location-Independent *k*-connected Scheme in Wireless Sensor Networks*

Xiaofeng Liu[1],   Liusheng Huang[1],   Wenbo Shi[1],   Hongli Xu [1]

[1](Department of Computer Science and Technology, University of Science and Technology of China, Hefei 230027, China)
{xfliu, shiwenbo}@mail.ustc.edu.cn, {lshuang, xuhongli}@ustc.edu.cn

Abstract. The reliability of communication can be enhanced by increasing the network connectivity. Topology control and node sleep scheduling are used to reduce the energy consumption. This paper considers the problem of maintaining *k*-connectivity of WSN at minimum energy level while keeping only a subset of sensor nodes active to save energy. In our proposed scheme, each node is assumed to have multiple power levels and neighbor proximity not exact location information is adopted. Firstly the network partition is attained by power based clustering, and next nodes are divided into equivalent classes according to the role of data forwarding to different adjacent clusters. Then Node Scheduling and Power Adjustment (NSPA) algorithm selects a subset of nodes with different power levels to construct the local minimum energy graph while maintaining network connectivity. If the number of intra-cluster nodes which have adjacent clusters exceeds a certain threshold, *k*-NSPA is employed. Finally, a *k*-connected topology can be obtained. The simulation shows that our scheme can obtain the redundant nodes while maintaining network *k*-connected and it is more energy efficient compared with previous work.

Keywords: *k*-connectivity, power, clustering, equivalence

1   Introduction

Wireless sensor networks (WSNs) are being increasingly deployed in severe areas to monitor the environment [1, 2]. In these applications, the network must be maintained connected to ensure the availability and reliability of communication. WSNs systems always contain a large number of nodes, due to the distributed area is large and the environment around is complicated, changing the batteries to renew the node energy is not practical. So sensors are required to be redundantly deployed to accommodate unexpected failures and extend the network lifetime. If idle sensors are not asleep, then redundant node deployment does not necessarily improve the connected coverage time of the field. So an important challenge of WSNs is how to maintain network communication connectivity and utilize the redundant nodes to extend the

* Supported by the National Grand Fundamental Research 973 Programs of China under Grant No.2006CB303006, No.2007CB316505 and the Knowledge Innovation Project of the Chinese Academy of Sciences.

Please use the following format when citing this chapter:

Liu, X., et al., 2008, in IFIP International Federation for Information Processing, Volume 264; Wireless Sensor and Actor Networks II; Ali Miri; (Boston: Springer), pp. 257–268.

network lifetime. An important frequently addressed objective is to determine a minimal number of working sensors required to maintain the connectivity. Therefore, the network topology should be controlled by selecting a subset of nodes to actively monitor the field and putting the rest nodes into sleep. The existing researches on node scheduling [10-15] are mainly address the scheduling mechanism to maintain the connectivity under maximal power. There are many researches [3-9] on topology control to gain energy efficiency. We can combine node scheduling and topology control to realize energy efficiency while maintaining connectivity. Most proposed protocols for topology control and node scheduling assume that nodes can estimate their locations or at least the directions of their neighbors (e.g., [14, 15]). In this work, we focus on applications in which location is unnecessary and possibly infeasible. Thus, we need redundancy nodes check that are based on neighbor proximity rather than on locations. Location independent connected coverage problem researches [16, 17] aimed at the relationship of neighbor set associated the node to achieve connected coverage. They are mainly based on the maximal power without energy efficient topology control, so they may not reach the energy efficient objective. Therefore, it is an appealing idea to make use of the neighbor proximity together with power topology control to provide network connectivity.

The contribution of our work is introducing an energy efficient location-independent k-connected scheme by multi-level power topology control and node scheduling relying on node equivalence relationship. It saves energy by turning off the forwarding units of nodes which are redundant for the desired connectivity by multi-level power control. The benefits of the proposed technique are twofold. First, the node clustering technology is adopted to get one hop connected clusters and its adjacent clusters; by the equivalent role they act in data forwarding to the adjacent clusters, the intra-cluster nodes are divided into some equivalent classes and the redundant nodes are found while maintaining required connectivity. Second, the energy efficient topology is constructed while maintaining k-connectivity by multi-level power control without knowing node locations.

The rest of this paper is organized as follows. Section 2 gives an overview of related work. In Section 3, the proposed clustering scheme and energy efficient k-connected scheme are presented. Section 4 discusses the performance evaluation of the proposed scheme. Concluding remarks of our research are made in Section 5.

2   Related Work

Power control is quite a complicated problem. Kirousis simplifies it as a range assignment (RA) problem [3]. The objective of RA is to minimize the network sending power (minimize the total power of all the nodes) while maintaining network connected. RA problem could be solved in $O(n^4)$ polynomial time in one dimension, but in two dimensions [4] and three dimensions [3], it is NP hard. Thus trying to finding the best result of the power control is not practical, but the feasible result can be given in special application level. The existing work is mainly based on reducing the sending power to prolong the network life.

In COMPOW [5], all the sensor nodes adopt the same sending power to minimize the network power while keeping connectivity. When the nodes are distributed uniformly, this algorithm shows good performance. But isolated nodes will lead to larger sending power for all the nodes. LUSTERPOW [6] improves that. It uses the minimal power not the common power to forward to next hop. CBTC [7] is based on direction while keeping connectivity. This kinds of algorithms based on direction require reliable direction information, thus need to be equipped with several antennae and high capability. DRNG and DLMST [8] are typical algorithms based on adjacent graph. The power control based on adjacent graph requires accurate location information. XTC [9] algorithm uses the receiving power intensity instead of the distance metric in RNG. It does not require the location information, but not practical in some extent.

The sensor nodes coming into sleep status can save energy consumption. Kumar presents a simple sleep scheduling algorithm RIS [10], in which the node independently decides itself to go to sleep or not in some probability at the beginning of every period. Obviously, RIS requires strict time synchronization. MSNL [11] proposed by Berman views the node sleep scheduling problem as a maximal the network lifetime problem with coverage restricted. It also needs accurate location information and there may be many nodes coming into sleep at the same time. LDAS [12] without requiring location information presented by Wu is based on the fractional redundancy to schedule the nodes, offering the coverage statistical guarantee. HEED [13] is a typical Level based node scheduling algorithm, which realizes clustering by periodic iteration according to the residual energy and communication cost of intra-cluster. HEED adopts average minimum reach power (AMRP) as the communication cost when selecting cluster head, which offers a uniform clustering mechanism not like other clustering schemes. HEED realizes clustering independent of the network size and it integrates the lifetime, extendibility and load-balance into consideration ignoring the node distribution. It is executed independent of synchronization, but no synchronization leads to bad clustering quality.

3   Network Partition and k-Connected Scheme

We first introduce our system assumptions and then describe the proposed scheme for determining node redundancy. At last we construct k-connected energy efficient topology.

3.1   System Model

The assumptions are listed as follows:

1) Nodes are randomly and redundantly deployed. They have similar batteries and energy consumption rates. And they have multi-level power which can be used to obtain the neighbor proximity achieved by checking whether the receiving power is in some bound or not, which means that the neighbor nodes will be sorted into multi-level sets according to the power levels. The minimal power level is the required sending power for communication.

2) Nodes have omni-directional antennae and do not possess localization capability. Thus, node locations and relative directions of neighbors cannot be estimated. We adopt the antenna model as follows: Assume sensors transmit with a power P_s, let the signal attenuation over space be proportional to some exponent γ of the distance d between two nodes, times the antenna directivity gain G, ($G = 1$ for omni-directional antennae), that is, $P_r / P_s = lG^2 d^{-\gamma}$, with $2 \leq \gamma \leq 5$, where c denotes a proportionality constant, and P_r denotes the minimum required receiving power for communication.

Given connected graph G, try to construct a connected energy minimum graph H. For each node u, it will be assigned transmission power level, and our objective is to minimize transmission power while maintaining the network connectivity. Therefore, we need to construct H with as few nodes as possible to consume as low power as possible. We offer a distributed approximate algorithm to address this objective.

3.2   Network Partition and Equivalence Classification

In this subsection, we will introduce a distributed network clustering scheme to provide one hop connected clusters and the sorting of the multi-level neighbor nodes into several equivalent classes. The basic idea of clustering works as follows: First, virtual cluster head method is adopted. When neighbors initiate a link with the cluster, the reply will be answered by the virtual head. In the initiative stage, a few of nodes (called virtual cluster heads) spontaneously broadcast inviting messages to its neighbors at the lowest power level. Any neighbor node decides whether or not to join the cluster dominated set by that sponsor according to the receiving power. Then this set will send messages to its neighbors following a sequence of power levels in an ascending order. In this way, the set is enlarged bigger and bigger until each node has an ascription at last, then the clustering is finished.

This strategy contains 3 kinds of messages: 1) "Hello" message which is used in initiating a cluster; 2) "Probe" message that is to ask for an ascription; 3) "Reply" message used for notifying others about its own response.

A.   Multi-level Power Control Based Clustering and Adjacent Clusters Identification

1) Initially, some sponsors start to establish the clusters. Each of them appoints itself as the virtual head and makes an identification of this cluster.

At first, the virtual head broadcasts "Hello" message at the lowest power level P_0, see the Table 1 (Formalization and Description). Referred to the typical antenna model as mentioned in section 3.1, the receiving power can be described as a function of sending power and distance, which is formulized as $P_r = f(P_s, x)$. Each neighbor whose receiving power is in $[f(P_0, R_0 / 2), f(P_0, 0))$ will join into the cluster of this sponsor and CDS_0 is obtained. Then all nodes in CDS_0 will broadcast "Hello" message at sending power level P_1 (as shown in Fig.1). When the power level of all nodes in the cluster is increased from P_{i-1} to P_i, the range of new cluster will cover the range of original one and they follow an equation as $R_{i-1} + 2x_i = R_i$, so $x_i = (R_i - R_{i-1}) / 2$. By the formula $P_r = f(P_s, x)$, node whose receiving power is in $[f(p_i, x_i + R_{i-1}), f(P_i, R_{i-1})]$

declares to join in this cluster and set the ascription cluster ID. Then CDS_{i-1} is updated to CDS_i.

Table 1.   Formalization and Description

NOTATION	Descriptions
m	Total number of levels adjustable for power
P_i, R_i $(i = 0,1,\cdots,m-1)$	Sending power and Maximal communication range at P_i
CDS_i $(i = 0,1,\cdots,m-1)$	Cluster dominating set at P_i
CDS_{ij} $\begin{array}{l}(i = 0,1,\cdots,m-2, \\ j = i+1,\cdots,m-1)\end{array}$	Maximal communication coverage set of CDS_i
x_i $(i = 0,1,\cdots,m-1)$	Outspread range of CDS_i at power P_i
CDS, CDS_{max}	Final cluster dominated set and coverage set
C_i $(i = 1,\cdots,k)$	Adjacent cluster i
G_i $(i = 1,\cdots,k+1)$	Equivalent nodes set of a cluster

All nodes in CDS_i send messages to the neighbors at power level P_j, $i < j < m$, then CDS_{ij} is established. That is to say, $CDS_{ij} - CDS_i$ contains the nodes which are in the communication coverage of CDS_i at P_j but not belong to CDS_i. $CDS_{ij} - CDS_i$ means the neighbor set of CDS_i based on power P_j control. Assume final cluster dominated set and coverage set as CDS and CDS_{max}. CDS relays messages via nodes in $CDS_{max} - CDS$, so $CDS_{max} - CDS$ consists of nodes those relay packets to the adjacent clusters. Each node in $CDS_{max} - CDS$ will save this cluster ID as one of its adjacent clusters.

2) Each node checks up its cluster ID. If it finds that it does not belong to any cluster, it will broadcast the "Probe" message at the max power level P_{max}. After a certain period, if no "Reply" message is received, it will choose a random number as its cluster ID and return to 1).

3) When a node receives a "Probe" message, it will judge itself whether a virtual cluster head. If it is and its receiving power is in $[f(p_i, x_i + R_{i-1})P_m / P_i, f(P_i, R_{i-1}) P_m / P_i]$, then it will respond to the sender. Otherwise, the "Probe" message will be discarded.

During the process of clustering, if some clusters have nodes in common, the cluster which has the most nodes will continue, while others have to withdraw. So nodes in the common area are notified their cluster ID as the same as the biggest cluster. Nodes in smaller clusters not belonging to intersection set cluster ID as Null.

B. Equivalence Classification

After clustering, according to the forwarding communication function to different adjacent clusters, nodes are divided into many equivalent groups (as shown in Fig. 2). Assume there are n nodes in cluster A, and cluster A has ω adjacent clusters, noted as adjacent cluster C_j, $1 \le j \le \omega$. Intra-cluster nodes are divided into $\omega + 1$ groups like

$\{G_1, G_2, \cdots, G_\omega, G_{\omega+1}\}$. Nodes in the same group are equivalent according to forwarding role, noted as *equivalent property*. If node u in cluster A is in the coverage of adjacent cluster C_j at the power level of C_j, $1 \le j \le \omega$, i.e. $u \in CDS_{\max}(j) - CDS(j)$, then $u \in G_j$, all the nodes in G_j are equivalent as a role to relay packets to cluster C_j. If node u in cluster A is not in the range of any adjacent cluster, i.e. $u \notin CDS_{\max}(j) - CDS(j)$, $\forall j \in \{1, 2, \cdots, \omega\}$, then $u \in G_{\omega+1}$, that is to say, node u will not be chosen as a relay node by any adjacent cluster. The nodes for relaying are all in $G_1, G_2, \cdots, G_\omega$.

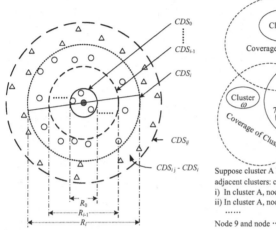

Fig.1. Some correlative sets in clustering.

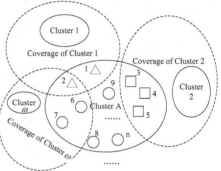

Suppose cluster A has nodes 1, node 2, ⋯⋯ and node n and has adjacent clusters: cluster 1, cluster 2, ⋯⋯ , and cluster ω.
i) In cluster A, node 1 and node 2 are equivalent to cluster 1.
ii) In cluster A, node 3, node 4 and node 5 are equivalent to cluster 2.
⋯⋯
Node 9 and node ⋯ are not in the coverage of any adjacent cluster.

Fig. 2. The equivalent classes after node clustering

3.3   Node Scheduling

From the process of clustering, we can find that any node in the network will be classified into a cluster, and any two clusters do not have a node in common. Considering the process of cluster establishment, we can obtain that the intra-cluster nodes are one hop connected by using a method of induction. The new cluster dominating set is extended by extending transmission range to R_i. Therefore they will be in each other's range at P_i. Now we consider some problems as follows:

A. The Node Redundancy Problem

Any node in intra-cluster, suppose $u \in G_i$, $1 \le i \le \omega$, they are equivalent to relay messages to their adjacent cluster C_j and in the transmission range of each other. If $|G_i| > 1$, it shows that there are redundant nodes in G_i, so just turn off the redundant nodes and wake up them by node sleeping schedule later. Look at such an occasion where node u belongs to a few of equivalent groups, for example, $u \in G_i \cap G_j \cap \cdots \cap G_k$. If u is turned on, all nodes like $v \in G_i \cup G_j \cup \cdots \cup G_k - u$ will be considered as redundant nodes, so turn them off. Denote adjacent clusters number of node u as index(u), i.e.

equivalent properties. What is optimal for energy saving is to turn on the low power level nodes those have the most equivalent properties.

B. The Problem of Connectivity between Adjacent Clusters

If a node u in cluster A is selected to be turned on, moreover $u \in G_i$, suppose u in the transmission range of adjacent cluster C_i at power level P_j, there will exist a node v, $v \in C_i$, such that node u and v are connected with each other at sending power level P_j. If node v in adjacent cluster is turned on, nodes which are in the same equivalent group as v will be regarded as redundant nodes.

C. The Problem of Power Asymmetry between Adjacent Clusters

After node clustering, if two adjacent clusters A and B are in different sending power levels P_A and P_B, and there exist two nodes $u \in A$, $v \in B$, such that u and v can hear each other at power level P_A or power level P_B, then choose the bigger bound $\max\{P_A, P_B\}$ to guarantee that intra-cluster nodes can connect with each other.

Node Scheduling and Power Adjustment Algorithm (NSPA)
Input: A set of sensor nodes with different power levels which are obtained by the network partition scheme and some clustering information, in which adjacent index denotes the equivalent properties of the node.
Output: A connected energy-efficient graph H.

NSPA Process //running by active nodes
    while (there are clusters without running this process)
        Sort the nodes according to adjacent index descending, noted as link L.
        Put the adjacent clusters' ID in the set S.
// find minimum nodes set connected with adjacent clusters
S1:   **for** the node u in the descending link L
        if (u.adjacent_cluster_ID element not in S)
            Delete the node from the link L and turn off u
        else
            Delete element in u.adjacent_cluster_ID from S
        $L = L \to next$
    if (S is Null)   **break**;
    for each node v in the link L
    Delete the node from the link L and turn off the node v
// search the nodes in adjacent clusters connected with L
    for each node u in the link $L \to head$
        for each cluster c in u.adjacent_cluster_ID
            Search v in c connected with u with the maximal adjacent index
            Set max$\{u$.power-level, v.power-level$\}$ for u and v
            Sort the nodes except v by adjacent index in c descending, link C.
            Put adjacent clusters' ID of c except the current cluster in the set S.
            Delete element in v.adjacent_cluster_ID from S
            View the cluster c as the current cluster and C as L, **goto** S1
    Return the active node set

Theorem 1: Given a redundantly connected graph G, topology H is connected after NSPA algorithm.

Proof: The final topology of the network consists of many clusters and each cluster has its own adjacent clusters. If there exists one cluster does not have any adjacent cluster, then the original network is not connected. Suppose two adjacent clusters A and B, there exists a node u, $u \in CDS(B)$, and $u \in CDS_{max}(A) - CDS(A)$. Node u in cluster B is a forward node for cluster A. If no such a node u exists, it shows that there is a cluster not in the transmission range of the other, so they are not adjacent clusters. Any cluster will connect to CDS through relay nodes in CDS_{max} CDS. Because any two adjacent clusters are connected, the network topology cluster based is connected. Furthermore, those intra-cluster nodes can hear each other within one hop, so they are connected. By NSPA which turns off some redundant nodes, the cluster is in the communication coverage of any intra-cluster node and keeps the ability to relay messages to its adjacent clusters. Connectivity exists between any two of the clusters, so after node scheduling, the topology of network H is also connected. $\square$

3.4   *k*-connected Algorithm

During clustering, we add an exit condition into the clustering process that the number of nodes which has adjacent clusters in this cluster is no less than k. And we try to find the minimum power consumption for active nodes set while keeping the graph k-connected. After running Alg. NSPA, the localized active set is obtained. Suppose the set of intra-cluster nodes awake is X, if $|X| \geq k$, the connectivity requirement is satisfied as proved in theorem 2, otherwise, in the cluster, choose at least $k - |X|$ redundant nodes which have adjacent clusters and turn them on to meet the requirement of k-connectivity. Let Z denote the node set which is turned on. Assume E represents the nodes set which is turned on in adjacent clusters connected with Z, satisfying $|E| \geq k - |X|$. In order to reduce energy as much as possible, we wake up the nodes in adjacent clusters whose power is no more than the redundant nodes in current cluster. Meanwhile, wake up the nodes which are connected with them in adjacent clusters.

k-connected Node Scheduling and Power Adjustment Algorithm (k-NSPA)
Input: The output of NSPA and the redundant nodes set of every cluster
Output: A k-connected energy-efficient graph H.

k-NSPA Process
    while (there are clusters without running this process)
        if ($|X| \geq k$)   **return** X
        Sort redundant nodes by the adjacent cluster power level ascending, link L.
        for node u in the descending link L
            $z = z + 1$ // initializing $z = 0$
            Turn on u and the appropriate nodes in E
            if ($z \geq k - |X|$ and $|E| \geq k - |X|$)   **break**;
            $L = L \rightarrow next$
    Return the active nodes set

Theorem 2: After k-NSPA process, the network H is k-connected.

Proof: Given a connected graph G, after the classification of equivalent groups, $\forall u \in G$ can hear any intra-cluster node within one hop. The number of intra-cluster nodes which have adjacent clusters is no less than k, so the number of intra-cluster nodes is no less than k. After running k-NSPA algorithm, the graph H obtained is a topology with some clusters. Any set with a number of $k-1$ nodes in the graph, $K = \{u_i \mid u_i \in H, 1 \le i \le k-1\}$, test the connectivity of $H-K$. In cluster A, assume X represents the active nodes set after Alg. NSPA, Z denotes redundant nodes set which will be turned on after Alg. k-NSPA. Suppose there are n nodes from K in cluster A, $n \le k-1$, it is obvious that $|X+Z-K| \ge 1$. If $n = k-1$, $\forall u \in X+Z-K$, if $u \in X$, then there exist some nodes in adjacent clusters which are connected to u; if $u \in Z$, while E represents the redundant nodes set which are turned on in adjacent clusters connected to Z, $|E| \ge k - |X|$, then there exist some nodes in adjacent clusters which are connected to u. If $n < k-1$, there exist nodes removed from other clusters, if these nodes are not in the adjacent clusters of the current cluster, the connectivity will not be affected, assume these nodes are in the adjacent clusters, if $n = k-2$, then $|X+Z-K| \ge 2$, it means removing a node from the adjacent cluster, because of $|E| \ge k - |X|$, i.e. $|E| + |X| \ge k$, this node is connected with at most one node in $X+Z-K$, therefore, there must exist a node in $X+Z-K$ connected with adjacent clusters; when $n < k-2$, the same as above.

Above all, $H-K$ is still connected, so network H is k-connected. $\square$

4   Performance Analysis and Simulation

We evaluate the energy-efficient scheme through simulation. Assume that n nodes are randomly and redundantly distributed in 100×100 square meters field. We focus on a snapshot during network operation where the energy of each node can be scaled in multi-level power. We assign the maximal power level of the nodes such that the transmission range of the nodes is all 30 meters and there are 6 power levels with different transmission ranges. Nodes have omni-directional antennae and do not possess localization capability, adopting the antenna mode described in section 3.1. Then we focus on the following metrics: (1) size of the active set for maintaining connectivity, (2) size of the active set for maintaining k-connectivity, and (3) energy consumption in each case. For simplicity, we do not simulate the energy costs associated with control messages during the equivalent nodes network partition.

3.1   Connectivity and Active Set

We choose $n = 100$, denote the active set as VA, after running the NSPA algorithm, there is only 24 nodes being active to maintain connectivity with different power levels, all the nodes in the communication range of VA. In this topology, we run the k-NSPA algorithm to obtain the k-connected minimum energy graph with different sizes of active sets as shown in Fig.3. From the result we can obtain that only a subset of

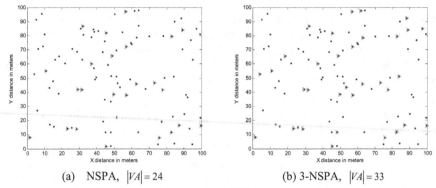

(a)   NSPA, $|VA|=24$ (b) 3-NSPA, $|VA|=33$

Fig. 3.   The active nodes set under k-NSPA algorithm with different k values.

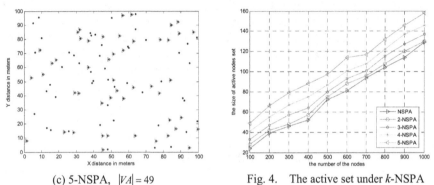

(c) 5-NSPA, $|VA|=49$ Fig. 4.   The active set under k-NSPA

Fig. 3. (c)   algorithm in different topology

the whole nodes are awake for forwarding, and $|VA|$ increases along with value k which refers to the network connectivity. This scheme is different from the previous research on topology control schemes which adjust the power of each node and the node scheduling scheme in which the node controls the status of itself to be active or sleep by judging the information of itself or being controlled by the cluster head. After NSPA algorithm, we select a minimum nodes in multi-power levels to get the local minimum energy graph while maintain network connected. Increasing the network connectivity, high communication reliability can be enhanced. After k-NSPA, a k-connected topology may be obtained.

In Fig.4, we give the active set size of the cases with different node number n and the different value k. Each point in our results is the average of 10 experiments of different random topologies. The algorithm offers a method of how to gain a k-connected graph from a connected graph by adding the condition the number of the nodes intra-cluster having adjacent cluster more than k. And awake the sleep nodes to enhance the network connectivity. This is a requisite condition to obtain the k-connected graph.

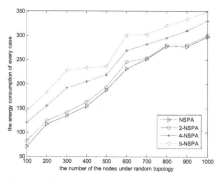

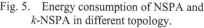

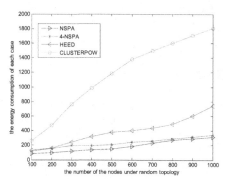

Fig. 5.   Energy consumption of NSPA and   Fig. 6.   Energy consumption of our scheme,
k-NSPA in different topology.   HEED and CLUSTERPOW.

3.2   Energy consumption

After the network partition and node scheduling, we can obtain the topology *H*. In our simulation, we assume the power level of the node as the proportion of energy consumption. The energy consumption is the total power levels of all the nodes. In the power level adjustment, select the nodes in these equivalent classes which can minimal the energy consumption locally. This algorithm is localized and distributed. We can obtain the localized minimum energy consumption, therefore achieving the approximate minimum energy graph which maintains the required network connectivity. The energy consumption increases as the network size and the network connectivity increase as shown in Fig.5. There is no node scheduling in COMPOW and LUSTERPOW. If the network is redundantly deployed, all the nodes keeping active leads to network disabled more early. So the energy consumption in these algorithms will be more than our scheme as shown in Fig.6. In the algorithm HEED, only node scheduling method is mentioned, there being no power control, in which all the nodes use the maximal power to send the packets. We compare the energy levels with ours, also shown in Fig.6. Obviously, our scheme is energy efficient.

5   Conclusion

This paper proposes a novel approach for topology control and node scheduling in the absence of location information. Our redundancy check relies on locally neighborhood proximity and equivalent classes by the association with the adjacent clusters. We incorporate our clustering scheme under multi-level power control and node scheduling into a distributed connected algorithm (NSPA). Next we add an exit condition in clustering, and then *k*-NSPA is employed, a *k*-connected topology may be obtained. Our scheme incurs low energy consumption and can significantly reduce the set of active nodes. Simulations show that the achieved energy efficiency by our scheme is significant to those achieved by some typical protocols. Although in the initial stages, such a location-independent *k*-connected scheme can be beneficial and effective for sensor applications that require high reliability and long lifetime of a specified sensor field.

References

1. T. He, S. Krishnamurthy, J.A. Stankovic, T.F. Abdelzaher, L. Luo, R. Stoleru, T. Yan, L. Gu, J. Hui, and B. Krogh. Energy-Efficient Surveillance System Using Wireless Sensor Networks. In: ACM MobiSys, June 2004, pages 270--283.

2. R. Szewczyk, A. Mainwaring, J. Polastre, J. Anderson, and D. Culler. An analysis of a large scale habitat monitoring application. In: Proc. 2nd ACM Conference on Embedded Networked Sensor Systems (SenSys '04), Nov. 2004, pages 214--226.

3. Kirousis LM, Kranakis E, Krizanc D, Pelc A. Power consumption in packet radio networks. Theoretical Computer Science, 2000, 243(1-2):289−305.

4. Clementi A, Penna P, Silvestri R. On the power assignment problem in radio networks. ACM/Kluwer Mobile Networks and Applications (MONET), 2004,9(2):125−140.

5. Narayanaswamy S, Kawadia V, Sreenivas RS, Kumar PR. Power control in ad-hoc networks: Theory, architecture, algorithm and implementation of the COMPOW protocol. In: Proc. of the European Wireless Conf. Florence, 2002. 156−162.

6. Kawadia V, Kumar PR. Power control and clustering in ad-hoc networks. In: Mitchell K, ed. Proc. of the IEEE Conf. on Computer Communications (INFOCOM). New York: IEEE Press, 2003. 459−469.

7. Li L, Halpern JY, Bahl P, Wang YM, Wattenhofer R. A cone-based distributed topology control algorithm for wireless multi-hop networks. IEEE/ACM Trans. on Networking, 2005,13(1):147−159.

8. Bahramgiri M, Hajiaghayi MT, Mirrokni VS. Fault-Tolerant and 3-dimensional distributed topology control algorithms in wireless multihop networks. In: Proc. of the IEEE Int'l Conf. on Computer Communications and Networks (ICCCN). 2002. 392−397.

9. Wattenhofer R, Zollinger A. XTC: A practical topology control algorithm for ad-hoc networks. In: Panda DK, Duato J, Stunkel C, eds. Proc. of the Int'l Parallel and Distributed Processing Symp. (IPDPS). New Mexico: IEEE Press, 2004. 216−223.

10. Kumar S, Lai TH, Balogh J. On k-coverage in a mostly sleeping sensor network. In: Haas ZJ, ed. Proc. of the ACM Int'l Conf. on Mobile Computing and Networking (MobiCom). New York: ACM Press, 2004. 144−158.

11. Berman P, Calinescu G, Shah C, Zelikovsky A. Efficient energy management in sensor networks. In: Xiao Y, Pan Y, eds. Proc. of the Ad Hoc and Sensor Networks, Series on Wireless Networks and Mobile Computing. New York: Nova Science Publishers, 2005.

12. Wu K, Gao Y, Li F, Xiao Y. Lightweight deployment-aware scheduling for wireless sensor networks. ACM/Kluwer Mobile Networks and Applications (MONET), 2005,10(6):837−852.

13. Younis O, fahmy S. HEED: A hybrid, energy-efficient, distributed clustering approach for ad hoc sensor networks. IEEE Trans. on Mobile Computing, 2004,3(4):660−669.

14. H. Zhang and J. C. Hou. Maintaining sensing coverage and connectivity in large sensor networks. Ad Hoc & Sensor Wireless Networks, 1:89−124, Jan. 2005.

15. X. Wang, G. Xing, Y. Zhang, C. Lu, R. Pless, and C. D. Gill, "Integrated Coverage and Connectivity Configuration in Wireless Sensor Networks," In: Proc. First ACM Conference on Embedded Networked Sensor Systems (SenSys'03), Nov 2003. New York.

16. Choudhury RR, Kravets R. Location-Independent coverage in wireless sensor networks. 2004. http://www.crhc.uiuc.edu/~croy/pubs

17. Ossama Younis. Marwan Krunz and Srinivasan Ramasubramanian. Coverage Without Location Information. IEEE International Conference on Network Protocols, ICNP 2007.

RAS: A Reliable Routing Protocol for Wireless Ad Hoc and Sensor Networks *

Imad Jawhar, Zouheir Trabelsi, and Jameela Al-Jaroodi

College of Information Technology
United Arab Emirates University
P.O. Box 17551, Al Ain, UAE
E-mail: {ijawhar, trabelsi,j.aljaroodi}@uaeu.ac.ae

Abstract. Wireless sensor and ad hoc networks are gaining a lot of attention in research lately due to their importance in enabling mobile wireless nodes to communicate without any predetermined infrastructure. Routing protocol in wireless sensor and ad hoc networks discover a multi-hop route between source and destination nodes. This paper presents RAS: a Reliable routing protocol for wireless Ad hoc and Sensor networks. In the RAS protocol, increased reliability is achieved by the maintenance of a reliability factor by the nodes. The value of this factor is increased when nodes participate successfully in data transmissions. This is determined through the use of positive and passive acknowledgements. During the path discovery process, an intermediate node only extends the request message to nodes that have a minimal reliability factor which is specified by the source. Additional optimizations are included in order to increase the efficiency and performance of the network.

Keywords: Mobile ad hoc networks (MANETs), wireless sensor networks (WSNs), reliability, quality of service (QoS), routing.

1 Introduction

The lack of a fixed topology and central control in mobile ad hoc and sensor networks poses a great challenge to the routing process in this environment. Particularly, when the issue of trust is in question. Routing protocols designed for ad hoc networks such as the Dynamic Source Routing protocol (DSR) [16], Ad hoc On-Demand Distance Vector (AODV) protocol [17], Temporally Ordered Routing Algorithm (TORA) [14], and many others [2][4][6][7][8][9][13][19][21][22] work very well under certain conditions where nodes are trusted, they all behave correctly and there are no intruders or malicious attacks on the network. However, we run into problems when we consider the reality that not all nodes will be cooperative and there are no guarantees that any of the nodes will be malicious. Routing protocols were investigated and modified and new protocols were introduced to enhance the routing process in MANETs. Many of these protocols provide some solution to parts of the problem.

To start, in [20] the authors argue that TCP is not suitable for ad hoc networks and propose a new transport layer routing protocol ATP. This enforces our approach to

* This work was supported in part by UAEU Research grant 08-03-9-11/07.

Please use the following format when citing this chapter:

Jawhar, I., Trabelsi, Z. and Al-Jaroodi, J., 2008, in IFIP International Federation for Information Processing, Volume 264; Wireless Sensor and Actor Networks II; Ali Miri; (Boston: Springer), pp. 269–279.

providing routing at a higher level and allowing the applications to take control of the process. Furthermore, the utilization of middleware to provide this type of functionality is another viable approach. The study in [5] provides an overview of possible middleware schemes and how some provide feasible solution to the ad hoc routing issues. One way to increase reliability is by using regenerating nodes [11]. This approach to ad hoc routing increases the reliability of packet delivery by allowing intermediate nodes on the routing path to reconstruct packets and send them to the destination in case of a problem. This reduces the traffic between the source and destination and speeds up the recovery of lost packets. However, this scheme assumes regenerating nodes are always available and willing to do the job and does not account for selfish nodes. Yet in many cases those nodes may not be able to do that because of their mobility, low resources, high traffic or plainly because they are malicious. Another approach is using a distributed multi-path DSR protocol for MANETs to improve QoS support for end-to-end reliability [10]. The protocol forwards outgoing packets along multiple paths based on specified end-to-end reliability requirements. As a result the packet will have better chances of arriving at the destination; however, this introduces high traffic volumes on the links due to the duplicate packets. An enhancement to DSR called DSR with Connection-Aware Link-state Exchange aNd DiffERentiation is introduced [3] to effectively collect and disseminate neighbor link states to nodes which may potentially use them. Neighbors exchange link-state information as soon as a connection is established using piggy-backed messages. This information speeds-up route discovery in case of link failures. In addition fidelity is used to assess the cost of a link, so when a new/alternate route has to be computed, the level of fidelity is figured into the link cost. In another approach, a reputation-based system built on the use of state model [1] is introduced to detect selfish nodes and encourage them to be cooperative by providing benefits. The techniques also handle suspicious nodes that may behave selfishly and encourage them not to. While the authors in [12] combine the reputation-based technique and the virtual currency to enforce cooperation between nodes. The combined approach takes advantage of the fairness of the virtual currency mechanism, while maintaining high cooperation levels based on the reputation approach. Another issue being investigated is trust. The authors in [18] introduce a mechanism to establish trust between nodes based on their reputation. During the network lifetime, nodes gain reputation value as they behave well and that allows other nodes to assess the reputation of their neighbors and the nodes to be selected for the routes needed. This approach is mainly useful to detect nodes that maliciously drop packets and allow nodes to avoid these malicious nodes. Furthermore, the authors in [15] provide mechanisms to detect attacks on the ad hoc network from selfish nodes and introduce a low-cost scheme to inform others about the accusations. In addition, the approach provides an inference scheme to back the accusations and allow nodes to make informed decisions on how to deal with selfish nodes. Many of the approaches discussed here and in other research articles target one main goal which is enhancing the routing process and maintaining reliable communication over multi-hop ad hoc networks. Our approach provides reliable routing based on a reliability factor established and maintained by the nodes in the network. The reliability factor is maintained by the nodes and is updated based on the successful participation of nodes in previous data transmissions. Our protocol is based on the

DSR routing protocol, and is on-demand. In addition, it is a distributed protocol. Unlike other protocols such as the link-state-based ones, each node only has to maintain topology and reliability information about its immediate neighbors and not the entire network. These characteristics enhance the scalability and performance of our proposed protocol.

The remainder of the paper is organized as follows. Section 2 presents the RAS routing protocol, along with the associated data structures, messages, and algorithms. Simplified and detailed examples are also provided. Section 3 discusses additional security issues and related future work in ad hoc and sensor networks. Finally, the last section concludes the paper.

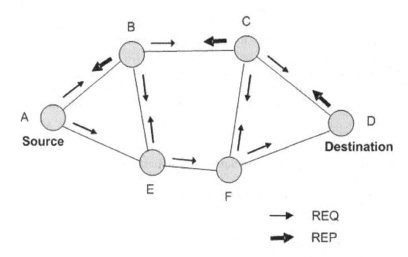

Fig. 1. A simplified example of the route discovery process.

2   The RAS Routing Protocol

This section describes the RAS routing protocol. It starts with a brief description of the DSR-based routing process along with a simplified example. Next, the RAS routing protocol data structures, messages, parameters, and algorithms are described along with a detailed example that illustrates the routing process. The maintenance of the reliability factor and data transmission process are described later.

2.1   A simplified overview of the DSR-based routing process

The protocol that is presented in this paper is based on the Dynamic Source Routing (DSR) protocol. It is on-demand which makes it more scalable. Nodes do not need to keep information about the entire topology and routes are only discovered as the need arises. When a source node s wants to send data to another destination node d which is not within its transmission range. It will try to discover a multi-hop path to it. In order to do that, node s broadcasts a request (REQ) message to all of its neighbors. Each of the neighbors adds its ID to the accumulating path in the message and in turn forwards it to all of its neighbors. This process continues until the REQ message reaches the destination, which then unicasts a reply (REP) message back to the source. Upon receiving the REP message the source updates its routing table and starts the data transmission process.

A simplified example of the route discovery process is shown in Figure 1. In the figure, node A is a source node that needs to send data to a destination node D. Node A checks its routing table and realizes that it does not have a path to D. Consequently, node A starts a route discovery process by broadcasting a REQ message to its neighbors B and E. Nodes B and E, not having processed a REQ with the (s, d, ID) tuple, check the information in the REQ message and realize they are not the destination and that they do not have a path to D. Each of them forwards the REQ message to its neighbors except for the node from which they received the request (i.e. A). This process continues until the REQ message reaches the destination D. Node D then unicasts an REP message with the discovered path $A - B - C - D$ to A. Upon receiving the REP message, node A updates its routing table and starts the data transmission process along the discovered path.

2.2   The request message and associated data structures

In our protocol, the source sends the request message REQ (s, d, id, x, r_{min}, r_{max}, r_{cum}, PATH, NH,) which contains the following fields:

1. s: ID of the source node.
2. d: ID of the destination node.
3. ID: Message ID. The (s, d, ID) triple is therefore unique for every REQ message and is used to prevent looping.
4. x: The node ID of the host that is forwarding this REQ message.
5. r_{min}: The minimum value for the reliability factor required in the path from s to d.
6. r_{cum}: The cumulative reliability factor the the path that is being discovered.
7. $PATH$: It contains the accumulated list of hosts, that the REQ message has passed through.
8. NH: It contains the next hop information. If node x is forwarding this REQ message, then NH contains a list of the next hop host candidates that satisfy the reliability requirements for the path that is being discovered.

2.3   Application specific parameters

In addition to the parameters specified above, the application layer is able to specify certain parameters that are used in the route discovery and communication process. These parameters depend on quality of service (QoS) requirements of the application, and are the following:

- MAX_NH: Maximum number of nodes in the NH list. This parameter is intended to control the flooding of the REQ message during route discovery. If this number is infinity, then the process reverts to normal flooding. Otherwise, if this number is set to k, then only a maximum of k neighbors that satisfy the reliability requirements can be selected as next hop for the REQ message.
- $BACKUP\_PATHS$: This is the maximum number of backup paths that can be included in the routing table of the source node. If this number is 0, then only the most reliable path is selected by the source and stored in its routing table. Otherwise, if this number is more than 0, then the primary path along with up to $BACKUP\_PATHS$ discovered paths are stored in the routing table. The additional paths can be used as backup in case of failure of the primary path. The backup paths are stored in sorted order based on their path reliability factor.

2.4   An overview of RAS routing process

When an intermediate node y receives the REQ message from a node x, it checks if it already processed this message which is uniquely specified by the (s, d, ID) tuple. If it already processed this message, it drops it. This prevents looping. Otherwise, it checks if it is the destination indicated in the message. If it is not, it checks its routing table for a path to the destination with the required minimum reliability factor. If such a path exists, it appends it to the PATH list and unicasts a reply REP(s, d, ID, x, PATH) message back to the destination. If a path to the destination does not exist in its routing table, it appends its ID to the PATH list in the REQ message and broadcasts the message to all of its neighbors excluding x. The PATH list in the message is an accumulated list of nodes that the REQ message has propagated through. This process continues until the REQ message arrives at the destination node d. At that time, d unicasts a REP message back to the source along the discovered path saved in the PATH list. When the source receives the REP message, it updates its routing table with this information and starts the data transmission process.

2.5   The algorithm at an intermediate node

Algorithm 1 presents the algorithm at an intermediate node in the RAS protocol. When an intermediate node y receives a REQ message from a node x, it sorts all of its 1-hop neighbors in descending order using their reliability factors $rf[z]$, where $z = 1..n_y$ are the 1-hop neighbors of y, and n_y is their total number. Then, the node builds the next hop list (NH) which includes a maximum of NH_MAX 1-hop neighbors of y with reliability factors higher than or equal to the minimum acceptable reliability factor, r_{min}. Afterwards, node y's reliability factor, r_y is added to the cumulative path reliability factor r_{cum} which is included in the REQ message. Finally the REQ message is broadcast to all of y's 1-hop neighbors.

Upon the reception of the REQ message, each of the neighbors will in turn check if its ID is in the NH list that is included in the received REQ message. If its ID is not included then it drops the REQ message. Otherwise, it processes the REQ message and continues its propagation until the message reaches the destination node which will

---

Algorithm 1 The main algorithm at an intermediate node

When a node y receives a REQ message

    Let r_y be the reliability of y

    let r_{min} be the minimum acceptable reliability factor of the path

    let n_y be the number of 1-hop neighbors of y.

    let $rf[z]$ be the reliability factor of node z.

    let MAX_NH be the maximum number of 1-hop neighbors that can be included in the NH list

    $NH\_temp = \phi$

    Sort all 1-hop neighbors of y in descending order using $rf[z]$ as a key

    for $(z = 1; NH\_temp < MAX\_NH$ and $z \leq n_y; z = z + 1)$ **do**

        if $(rf[z] \geq r_{min})$ **then**

            Add z's ID to the NH_temp list

        end if

    end for

    $r_{cum} = r_{cum} + r_y$

    if $NH\_temp \neq \phi$ **then**

        let $PATH\_temp = PATH \mid y$

        broadcast $REQ(S, D, ID, r_{cum}, r_{min}, r_{max}, y$

        $PATH\_temp, NH\_temp)$ message

    end if

---

then unicast a replay (REP) message back to the source along the nodes collected in the PATH list to finish the path discovery process.

When and if the destination receives multiple REQ(s,d, ID) messages for the same path uniquely specified by the (s, d, ID) tuple, it can take one of the following two actions depending on the value of the application specific parameter MAX_PATHS that is defined earlier. If $BACKUP\_PATHS$ is equal to 0 then, the destination selects the path with the highest path reliability factor $PRF = r_{cum}/n$, where n is the number of intermediate nodes in the path (not including the source and destination nodes), and unicast a REP message back to the source. Otherwise, if $BACKUP\_PATHS > 0$, then the destination sends up to $(BACKUP\_PATHS + 1)$ REP messages (if that many have been discovered) back to the source which can use the path with the highest PRF as the primary path and the other paths as secondary paths which can be used as backup paths when the primary path breaks.

2.6   A detailed example

Figure 2 shows a detailed example that illustrates the route discovery process using the RAS routing protocol between the source node A and the destination node G. In this case, the required path to transmit the data has a minimum reliability factor $r_{min} = 5$, and $MAX\_NH = 2$. The source node, A, starts by sorting its 1-hop neighbors, nodes P, B, and S according to their respective reliability factors 10, 7, and 6. This information is contained in its reliability table which keeps updated information about the reliability factors of the node's 1-hop neighbors. All of A's neighbors meet the minimum requirement of 5. However, only nodes P and B are included in the NH list which can only

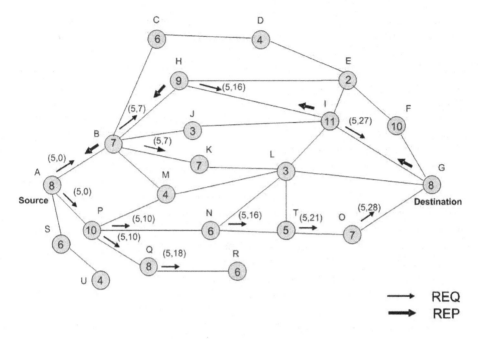

Fig. 2. A detailed example of the route discovery process.

contain a maximum of two 1-hop neighbors (for this path, $MAX\_NH = 2$). The NH list is included in the REQ message along with a cumulative path reliability factor r_{cum} of 0. Note that the source and destination nodes' reliability factors are not included in r_{cum} because the cumulative reliability factor is meant to measure the reliability of the intermediate nodes that will be used for data transfer along the discovered path. The source and destination nodes need to send and receive the data and must be a part of the final path by default regardless of their own reliability factors. Node A includes its ID in the PATH list contained in the REQ message and broadcasts it to its neighbors.

In turn, when node B receives the REQ message, it sorts its 1-hop neighbors nodes H, K, C, M, and J according to their respective reliability factors of 9, 7, 6, 4, and 3. Only nodes H, K, and C meet the minimum requirement of 5, and only nodes H, and K are included in the NH list, since $NH\_MAX = 2$. Subsequently, node B adds its ID to the PATH list included in the REQ message along with the constructed NH list and a cumulative reliability factor of $r_{cum} = 0 + 7 = 7$ (0 is the current cumulative reliability factor in the REQ message and 7 is the reliability of node B which will be broadcasting the REQ message). When the REQ message is received by the 1-hop neighbors, H, K, C, M, and J, each node will examine the NH list in the REQ message and check if its ID is included. Only the nodes whose id is included in the NH list will process the REQ message and try to propagate it further. The other nodes will simply drop the message. Using the same process, node H will further propagate the REQ message to node I with a cumulative factor $r_{cum} = 7 + 9 = 16$ (node E with the reliability factor 2 < 5 will not be included in node H's NH list). Node I will send the REQ message to node G with a cumulative reliability factor of $r_{cum} = 16 + 11 = 27$. Finally, the REQ message, arrives at the destination node G with the discovered path $A - B - H - I - G$ and a final cumulative reliability factor of 27.

Similarly to what node B did, node P, having received the REQ message from node A, will only include nodes N, and Q in its NH list along with a cumulative reliability factor of 10. The REQ message propagates from Q to R but does not go further since no further links exist out of R. The message also propagates from node N along nodes T and O to the destination node G with a final cumulative reliability factor of 28, and a discovered path which includes nodes $A - P - N - T - O - G$.

When the destination node G receives both REQ messages for the two discovered paths, it divides each path's final cumulative reliability factor with the number of nodes in the path to calculate the *normalized path reliability* factor $PRF = r_{cum}/n$. In this case, the destination node G determines that the discovered paths $A - B - H - I - G$, and $A - P - N - T - O - G$ have normalized reliability factors of $27/3 = 9$ and $32/4 = 8$ respectively. Therefore it chooses the more reliable path $A - B - H - I - G$ with the higher normalized path reliability factor of 9. It then unicasts a REP message back to the source node A along the discovered intermediate nodes in the PATH list to inform it of the discovered path. When node A receives the REP message, it updates its routing table with the discovered path and starts the data transmission process to node G. In our example, it is assumed that $BACKUP\_PATHS = 0$. However, if $BACKUP\_PATHS > 0$ then REP messages for both discovered paths would be sent back to the source. The latter will then use the path with the higher reliability factor as a primary path and saves the other discovered path in its routing table as backup path that would be used if and when the primary path fails later.

2.7   Maintenance of the reliability factor

The reliability factor of a particular node z is a measure of its past performance in being a part of successful data transmissions and it is maintained in the following fashion. At network initialization, all reliability factors are assigned an initial value $INIT\_REL\_FACTOR$. During normal network operation, once a path from the source to the destination is discovered, the source updates its routing table and starts data transmission along the discovered path through the intermediate nodes that belong to this path. There are several different algorithms that can be used to maintain the reliability factor:

Self maintenance of the reliability factor by the node: Each time a node successfully transmits a number of packets equal to $REL\_FACTOR\_RESOLUTION\_NUMBER$ to another node along the path it increases its reliability factor by 1. The $REL\_FACTOR\_RESOLUTION\_NUMBER$ is a constant set by the system administrator to make sure the actual reliability factor does not grow too large and is only incremented once for each predetermined number of successfully transmitted packets. This policy of updating the reliability factor by the node itself has minimal overhead and assumes no malicious nodes exist in the network. It is a policy designed to increase the reliability of the future discovered paths and not to combat malicious intentions by the nodes. Periodically, the reliability factor is automatically decremented by each node in order to keep this variable from overflowing while keeping the relative values of the reliability factors in the nodes consistent.

Maintenance of the reliability factor by the acknowledgements: Using this policy, the destination sends positive acknowledgments to the source. These acknowledgments are cumulative and can be piggy backed with data transmissions in the case of a two-way communication between two nodes in order to minimize overhead. the reliability factor is incremented when an intermediate node that transmitted participated in data transmission receives positive acknowledgements from the destination.

Maintenance of the reliability factor by 1-hop neighbors: Another strategy for maintenance of the reliability factor is that each node y maintains a reliability factor for each of its n neighbors z_1 - z_n. This set of reliability factors for each of its neighbors constitutes its a measure of the reliability of each of them in its view based on their past performance in successful data transmissions. The reliability factor rf[z] in node y is increased by 1 for each number of packets equal to $REL\_FACTOR\_RESOLUTION\_NUMBER$ forwarded by this neighbor. During the data transmission phase, when y forwards a data packet to a neighbor z which is not the destination and is a part of the discovered path for that session, it listens to the subsequent transmission by z to its successor node in the path. In this case, y is applying what is called a *passive acknowledgement* process, which is also a

Maintenance of the reliability factor using link-based-acknowledgement: If z did follow up and forward the packet to its successor node, then it is rewarded by increasing its reliability factor in y. Otherwise, if z did not forward the data as it should, then it is penalized by not increasing its reliability factor. Consequently, it is less likely to be selected by y for forwarding REQ packets during future path discoveries. Therefore, it is also less likely to be a part of future data transmission paths.

2.8 The data transmission process

Once the source receives the REP message from the destination, it updates its routing table with information of the newly discovered path and starts data transmission. During the data transmission process, positive acknowledgments are sent from the destination to confirm successful reception of packets. As indicated earlier, in order to save bandwidth and minimize overhead, acknowledgments are cumulative and can be piggy-backed with data messages that could be sent in the opposite direction from the destination to the source during a two-way communication session. The acknowledgements are used in the process of maintenance of the reliability factor at all of the nodes involved in the data transmission path. In addition to the acknowledgement process, the source node uses a time out timer named ACK_REC_TOT (acknowledgement reception time out timer), which is initialize at the beginning of the data transmission phase to the value $INIT\_ACK\_REC\_TOT$. The timer is refreshed each time an acknowledgment is received by the source. This indicates that the path to the destination is still valid. If the ACK_REC_TOT expires, then the source retransmits the unacknowledged packets. If the retransmitted packets are also not acknowledged, another retransmission attempt is made, until a maximum number of attempts is reached $MAX\_RETRANS\_ATTEMPTS$. At that time, the source assumes that the path is broken and starts another path discovery process.

3   Additional Security Issues and Future Work

This paper presented a reliable routing protocol for wireless ad hoc and sensor networks. However, malicious nodes in the network may affect the operation and efficiency of the proposed protocol. We identified four types of attacks that malicious nodes can generate. (1) The first type is related to the sniffing nodes. To do its attack, the malicious node redirects the selected traffic so that it can perform traffic sniffing. The second type is concerned with malicious nodes that attempt to attack the path discovery process. The malicious node does not extend the request message to nodes that have the minimal reliability factor specified by the source. Rather, the malicious node sends the request message to nodes with less reliable factor. So that, the source node will later use a non optimal path, hence damaging the efficiency and performance of the network. (3) The third type is concerned with malicious node that attempt to block any traffic that go through. Obviously, its reliability factor will decrease considerably, in the case where the factor is maintained by other nodes, and consequently no path will include that node. However, for a period of time, this attack may disturb the normal activities of the network. (4) The fourth type is concerned with a malicious node that, at the beginning, behaves as a very reliable node, in order to be included in the paths, and then the node can carry an attack. In such a situation, the network performance may be considerably affected since most traffic will go through the malicious nodes and consequently a denial of service attack can take place.

4   Conclusions

In this paper a reliable routing protocol for wireless mobile ad hoc and sensor networks was presented. The protocol provides increased reliability of communication by having intermediate nodes extend the REQ message only to the more reliable neighbor nodes using a reliability factor. The reliability factor is increased when nodes participate successfully during the data transmission process by using an acknowledgement mechanism. Positive and passive acknowledgement mechanisms are used in the maintenance of the reliability factor. Different parameters are used to optimize and enhance the routing process like reducing flooding during the path discovery phase, and the use of backup routes for more timing critical applications with increased priority. In the future, we intend to improve this protocol further by applying more techniques in optimizing the selection of the next-hop neighbors. In addition, we intend to further study, and analyze the performance of the protocol through simulation.

References

1. T. Anantvalee and J. Wu. Reputation-based system for encouraging the cooperation of nodes in mobile ad hoc networks. *Proc. of IEEE ICC*, 2007.
2. M. Barry and S. McGrath. QoS techniques in ad hoc networks. *Proc. of 1st International ANWIRE Workshop, Glasgow*, April 2003.
3. W.-P. Chen and J. C. Hou. Dynamic, ad-hoc source routing with connection-aware link-state exchange and differentiation. *Proc. IEEE Globecom*, November 2002.

4. S. De, S.K. Das, H. Wu, and C. Qiao. A resource efficient RT-QoS routing protocol for mobile ad hoc networks. *Wireless Personal Multimedia Communications, 2002. The 5th International Symposium on*, 1:257–261, 2002.

5. S. Hadim, J. Al-Jaroodi, and N. Mohamed. Trends in middleware for mobile ad hoc networks. *The Journal of Communications*, 1(4):11–21, July 2006.

6. Y. Hwang and P. Varshney. An adaptive QoS routing protocol with dispersity for ad-hoc networks. *System Sciences, 2003. Proceedings of the 36th Annual Hawaii International Conference on*, pages 302–311, January 2003.

7. I. Jawhar and J. Wu. Qos support in tdma-based mobile ad hoc networks. *The Journal of Computer Science and Technology (JCST)*, 20(6):797–910, November 2005.

8. I. Jawhar and J. Wu. Quality of sevice routing in mobile ad hoc networks. *Resource Management in Wireless Networking, M. Cardei, I. Cardei, and D. -Z. Du (eds.), Springer, Network Theory and Applications*, 16:365–400, 2005.

9. I. Jawhar and J. Wu. Race-free resource allocation for QoS support in wireless networks. *Ad Hoc and Sensor Wireless Networks: An International Journal*, 1(3):179–206, May 2005.

10. R. Leung, J. Liu, E. Poon, Ah-Lot. Chan, and B. Li. Mp-dsr: A qos-aware multi-path dynamic source routing protocol for wireless ad-hoc networks. *in Proc. 26th IEEE Conference on Local Computer Networks (LCN 2001)*, pages 132–141, 2001.

11. R. Ma and J. Ilow. Reliable multipath routing with fixed delays in manet using regenerating nodes. *in proc. 28th IEEE International Conference on Local Computer Networks (LNC'03), Bonn, Germany*, pages 719–725, October 2003.

12. Jamal N., Al-Karaki, and Ahmed E. Kamal. Enforcing cooperation in mobile ad hoc networks. *Springer Wireless Personal Communications*, 2006.

13. S. Nelakuditi, Z.-L. Zhang, R. P. Tsang, and D.H.C. Du. Adaptive proportional routing: a localized QoS routing approach. *Networking, IEEE/ACM Transactions on*, 10(6):790–804, December 2002.

14. V. D. Park and M. S. Corson. A highly adaptive distributed routing algorithm for mobile wireless networks. *INFOCOM '97. Sixteenth Annual Joint Conference of the IEEE Computer and Communications Societies. Proceedings IEEE*, 3:1405–1413, April 1997.

15. K. Paul and D. Westhoff. Context aware detection of selfish nodes in dsr based ad-hoc networks. *Proc. IEEE Semiannual Vehicular Technology Conference (VTC-2002)*, September 2002.

16. C. E. Perkins. *Ad Hoc Networking*. Addison-Wesley, Upper Saddle River, NJ, USA, 2001.

17. C. E. Perkins and E. M. Royer. Ad hoc on demand distance vector (AODV) routing. *Internet Draft*, August 1998.

18. Y. Rebahi, V. E. Mujica-V, and D. Sisalem. A reputation-based trust mechanism for ad hoc networks. *Proc. 10th IEEE Symposium on Computers and Communications (ISCC 2005), Spain*, pages 37–42, June 2005.

19. J. L. Sobrinho and A. S. Krishnakumar. Quality-of-service in ad hoc carrier sense multiple access wireless networks. *Selected Areas in Communications, IEEE Journal on*, 17(8):1353–1368, August 1999.

20. K. Sundaresan, V. Anantharaman, H-Y, Hsieh, and R. Sivakumar. A reliable transport protocol for ad hoc networks. *IEEE Transactions on Mobile Computing*, 4(6):588–603, Nov/Dec 2005.

21. H. Xiao, K. G. Lo, and K. C. Chua. A flexible quality of service model for mobile ad-hoc networks. *Proceedings of IEEE VTC2000-Spring, Tokyo*, May 2000.

22. Z. Ye, S. V. Krishnamurthy, and S. K. Tripathi. A framework for reliable routing in mobile ad hoc networks. *IEEE INFOCOM 2003*, 2003.

Author Index